Shepherd's Delight:

The Best of Tom Duncan

The author

'Tom Duncan' is the pen name of Rog Wood, journalist, author, and, of course, working farmer. His first book, *Magic Moments: Four Seasons on a Scottish Hill Farm*, published in 2004, was a critical and commercial success and has already been reprinted three times. Wood has written for a wide range of publications including *The Scotsman*, *The Herald*, *Yorkshire Post* and *The Scottish Farmer*. His column in the *Sunday Post* is still delighting readers after seventeen years. But farming is his first love and after thirty years working his 700-acre hill farm near Sanquhar he has no intention of retiring.

Shepherd's Delight:

The Best of Tom Duncan

Tom Duncan

Fort Publishing Ltd

First published in 2006 by Fort Publishing Ltd, Old Belmont House,
12 Robsland Avenue, Ayr, KA7 2RW

First photograph in plate section (of Rog Wood) courtesy of
Niall Robertson

Cover illustration by Andy Bridge

Graphics by Mark Blackadder

Typeset by Senga Fairgrieve

Printed by Bell and Bain Ltd, Glasgow

ISBN 10: 1-905769-00-8
ISBN 13: 978-1-905769-00-1

For my wife, Carmen

Contents

Preface

My father was a vet who bought a farm in later life, so I lived in Dumfries until I was eleven, when my father decided to become a full-time farmer and moved to the farm.

Growing up in the centre of a town meant that I had many friends living nearby that I could play with, but life on the farm was more isolated. I was the oldest of a family of four and my early teenage years were lonely and unhappy. For some reason my parents did not love me the way they loved my two sisters and younger brother. I was disruptive at school, constantly getting into trouble and a disappointment to my parents. Perhaps that was why they treated me differently, or maybe their harsh attitude towards me was the reason I became a difficult teenager. As they are now both dead I will never know, but I became bitter and resentful.

When I wasn't at school, or working on the farm helping my father, I spent a great deal of time by myself fishing for trout in nearby burns. During the winter months I liked to go out to the cattle sheds in the evenings to sweep-up hay in the feed passage and be with the cattle rather than spend time in the farmhouse and run the risk of yet another long-winded lecture from my father about what a disappointment I was to him.

After doing an HND at agricultural college I returned home to work on the farm, but my relationship with my father deteriorated. Our constant arguments began to upset my wife Carmen, who I had met at college and married in 1975, but we were fortunate to get a fresh start when I secured the tenancy of Auchentaggart in 1976 at the age of twenty-three. It was a badly run down, 700-acre upland farm on one of the Duke of Buccleuch's estates in

Dumfriesshire and I started the venture with very little capital of my own. My overdraft facilities were limited in those early days but local merchants and auctioneers were very supportive and gave me good credit facilities. It's amazing how many young men like me owe their start in farming to auction marts that allowed you to buy livestock on favourable credit terms.

My first few years were reasonably profitable allowing me to borrow more from the bank, and that was my downfall. I was extremely ambitious and soon set about radically improving the farm with the aid of the many generous grants that were available at the time. I drained the land, erected miles of new fences, rebuilt all the dilapidated dykes, ploughed, limed and reseeded the fields as well as putting up new sheds for cattle. Before long my overdraft had escalated to alarming levels and I ran the real risk of losing everything. To make matters worse, interest rates were rising at a time when farm profits were falling.

By this stage I was keeping about seventy beef cows and five hundred ewes, but, despite employing no staff, the farm struggled to break even and never made enough to provide a living for my young family.

As I struggled to make ends meet I became a regular and vociferous attendee of local farmer's union meetings. My outspoken and often controversial views were regularly reported in the press and that led to me becoming a regular columnist in *The Scottish Farmer*. It wasn't long before my editor tired of my outlandish style of writing and sacked me, but, shortly after, I was invited to write a regular column in *Farmers Weekly*. As has often happened in my life, when one door shuts another one opens.

Writing columns in various publications gave me the chance to earn badly needed extra income in the comfort of my own home after I had finished my work on the farm. A big breakthrough in my writing career came in 1989 when I started writing a weekly farming column in the *Sunday Post* under the name of

Tom Duncan. That was followed by my appointment as the Scottish agricultural correspondent for *Big Farm Weekly*, a popular publication that was circulated free to farmers throughout Britain. I now had a substantial income from writing that allowed me steadily to pay off my debts.

It's interesting to reflect that, although I eventually repaid all my loans, my younger brother wasn't as successful, despite inheriting the family farm.

My father was very proud of his farm and it is fair to say that he loved it more than anything else in his life. He started with nothing, worked hard to build up his veterinary practice, bought the farm cheap because, like mine, it was badly run down and through sheer hard work, thrift and a shrewd business sense built a very successful farming business.

Had he left it to his four children it probably would have been sold so that the money could be shared out equally, so he left it to my younger brother to avoid that. The decision was made easier because my brother was his favourite and he didn't get on with me. We eventually stopped speaking to each other some time before he died.

Like many farmers who develop a passion for their land, he wanted to be the founder of a farming dynasty where the farm would be handed down from father to son through successive generations; a kind of everlasting memorial to himself. He once said to me, 'You will find that you can't always be fair in farming' and I have often wondered if that was his way of preparing me for his decision to leave the farm to my younger brother. He should have reflected on verse 8, chapter 12 of the book of Ecclesiastes in the Holy Bible: 'Vanity of vanities, saith the Preacher, all is vanity.'

My brother was not a successful farmer and was eventually forced to sell the farm, abruptly stopping my father's dynasty in its tracks. That tragedy added to my deep sense of resentment. It had been difficult enough for me to come to terms with alienation from my parents, but watching my brother squander such a wonderful opportunity while my wife and I scrimped and saved to build up our own business from scratch made matters worse.

With hindsight, I was probably dealt a better hand in life than it first appears. Had I been handed a farm on a plate there is every chance that I would never have realised my full potential. Circumstances have forced me to try different things, to have many irons in the fire and above all else to develop my writing skills.

Another breakthrough in my life occurred in 1992, when I was about to turn forty, and was awarded a Nuffield farming scholarship. I travelled extensively throughout Eastern Europe, Italy and Spain studying the international trade in lambs. That scholarship was truly inspirational and gave me the confidence to win elections to become a director of the British Wool Marketing Board and an independent councillor on Dumfries and Galloway Council.

As my knowledge of the agricultural industry grew through my involvement in the various organisations associated with it, I became aware of the real need to explain farming matters to the general public. I now concentrate on writing for non-farmers, and, my first book, *Magic Moments* – also written under my *Sunday Post* pen name, Tom Duncan – was aimed at that wider audience, as well as practical farmers.

I believe that I have included enough humour and anecdotes to make this an enjoyable and entertaining read. I hope that this book will be as successful as *Magic Moments* and leave the readers with a better understanding of my rural way of life, farming and food production.

Rog Wood
Sanquhar
July 2006

1

My better half

There are many important ingredients needed to make a successful farmer. To my way of thinking good health, ability, enthusiasm, hard work and good luck rank among the more important. While good land or rich parents may be of some help, they are not essential and, in some cases, may even be a disadvantage.

Ranking above all other requirements for success is a good wife. There is a lot of truth in the old saying that behind every successful man there is a good woman.

Wives are important for many reasons and are an obvious requirement to start a family. Indeed, it is often the appearance of the next generation that spurs a farmer on to excellence. Even if it is not, there is no disputing that a happy, stable family is a powerful base for any farmer to work from.

Above all else, wives are companions for life, someone to share a joke with or be supportive at a time of disappointment or after a setback. Their constant presence in the decision-making process has a more profound effect on the development of a farmer's career than any host of bankers, accountants and advisers. A look from my wife can do more to influence that final decision than any budget or business plan.

It is only after working together and striving for something as a couple that you can really enjoy success. Working together in difficult times allows us to appreciate each other's strengths as well as understand our weaknesses. No marriage is a bed of roses; there will always be times when tempers are short and rows flare up. A strong marriage is one that learns to take these incidents in its stride.

For most farmers it's their wife who is the backbone of the business. At one time many led a life of drudgery: hand-milking cows; churning butter; cleaning the paraffin lamps that lit the byre; running a big farmhouse; preparing meals for family and workers.

That's all changed. Modern cattle sheds are lit by electricity, cows are milked mechanically and butter and cheese is no longer made on the farm. The tradition of providing meals for farm workers has died out. Kitchen floors are covered with vinyl that doesn't need to be vigorously scrubbed like the old grey flagstones.

Despite these changes for the better, most farmers' wives are as busy as ever. Many do the farm accounts and the mass of paperwork that floods a modern farm office. Then there's answering the phone and running errands such as fetching spare parts to repair a breakdown. Others drive tractors, or help with the livestock. Truth to tell, womenfolk are far better than men at rearing calves. That's one job where the woman's touch is important.

One of the biggest changes is the way many wives now earn extra money outside farming. Many smaller family farms don't make enough profit to live on nowadays. So some wives do farmhouse bed-and-breakfast, or use the car as a school taxi. Others do the farm work to let their husbands earn a living elsewhere. Quite a few cash in on their cooking skills with outside catering, such as wedding meals in the village hall or school canteen; the food is much better than hotel fare, and cheaper.

It's amazing how resourceful wives are at finding ways to make money: farm shops, antique shops, tearooms or pony-trekking ventures to name but a few. Then there are the more traditional ways of earning extra money like becoming a secretary, teacher or bank teller. Farmers have had to change their attitudes. Many of us have swallowed our pride and accepted we are no longer the sole breadwinner.

My father's generation was waited on hand and foot and got used to having their meals on the table on time. They couldn't even boil an egg! My generation on the other hand is much more domesticated. Although many of us still can't cook, we've learned to reheat the meals prepared for us. We take our boots off at the back door to keep the floors clean, or bring in the washing when it looks like rain. We have even learned to wash the dishes and leave the kitchen tidy!

Superwoman was the only one fit enough to have swapped places with my wife when our children were younger. Being a mother to three very active children is never easy, but it's more complicated when you live on a farm. As well as taking the youngsters to and from the school buses there were also a lot of extra journeys transporting them to their various activities and clubs.

As I don't employ anyone my wife has to help me on the farm. That can involve getting up at half-past five at lambing time to help me with the sheep and cattle. In the past, she then had to go back into the house at half-seven to get the children ready for school. After they were safely on the school bus she would rejoin me outside until they came home. Somehow, she fitted the children and housework into the busy routine of working with me.

In wet weather we both have to change into dry clothing several times a day. I have seen our oil-fired Raeburn cooker surrounded by chairs covered with wet overalls and coats hanging to dry. Often as not the washing machine is switched on every time we are in for a meal so that she can keep on top of the mountains of washing.

In between, she found time to prepare meals for the family and to keep the house tidy. Even that's not an easy task in a farmhouse with muddy wellies constantly at the back door. It's hard work and can get hectic but our system works reasonably well because I invested in the good years and have a modern, efficient farm.

One of the drawbacks to running the farm ourselves, though,

is arranging time off or getting cover if one of us is ill. A good example occurred one time I had to spend four days in Shetland on business. It was a quiet time of year on the farm and I was confident my wife would easily cope. Unfortunately, that trip coincided with torrential rain and gales. Shetland seemed to be the windiest, wettest place in the world that week. The ferry from Lerwick couldn't berth in Aberdeen, so the passengers had to spend extra time on the rough seas. Many fishing boats were tied up sheltering from the storm. Imagine my shock when I phoned my wife and discovered that the weather had been almost as bad at home.

A group of cows had been huddled in a corner during the night, sheltering from the blast, and had broken a wooden gate. My wife awoke the next morning to find that half of them had escaped.

Normally our neighbours would have helped her to round them up, but they were all away at an important sheep sale. As a result, she had to don her waterproofs and do the job herself. Hardly a pleasant task on a torrentially wet and windy day.

After a lot of running about she managed to get them back where they belonged and temporarily fixed the broken gate with string. Later that evening my daughter's boyfriend repaired the gate properly.

Having completed the task of rounding up the cattle she then realised that our steep, un-tarred farm road was in danger of being washed away by the floodwater. Armed with a pick and shovel, she had to clear the run-offs so that the water was diverted into the roadside ditch. It all made for a long, hard day in the worst of working conditions, but thanks to her hard graft and initiative I returned to a farm that was, as they say, shipshape and Bristol fashion.

A farmer's lot isn't an easy one, but it helps when his wife can more than pull her weight when he's away.

Mind you, sometimes the boot is on the other foot and I have to cope to let my wife away. That's always hectic and makes me appreciate how much work my so-efficient wife does. The biggest loss is that extra pair of hands for all the little livestock jobs.

For instance, on one of the occasions I was left to fend for myself a sick cow needed to be taken from the main cattle shed and put in a snug, straw-bedded pen to recuperate. It would normally have taken a few minutes for my wife and me to walk her down to the sick bay but, without her to help, that pesky cow ducked round every corner, enjoying having the upper hand for once.

The worst part of my wife being away is that it's a very lonely and long day working on the farm alone. My ordeal usually ends within a few days and often does me no harm but, in these tough times, many farmers' wives now recognise how important their company is at tea breaks and mealtimes. A smile and blether with a wife is often all it takes to prevent a worried farmer from becoming seriously depressed.

Without a doubt, a farmer's wife is the better half!

My neighbour has a house on his farm that used to be home to his shepherd. Times change and the shepherd was made redundant, so my neighbour's wife has furnished that house and lets it through an agency.

All manner of folk stay there. Families rent it for a holiday in the country whilst hillwalkers and bird-watchers use it as their base. Groups of canoeists use it in wintertime and brave a soaking in a chilly river because the angling club reckons canoes scare the fish and spoil their sport in the fishing season.

Recently my neighbour had an enquiry from an English couple needing a house for a couple of months. It seems that the husband had got a job in the area but didn't want the hassle of a bridging loan. He'd wisely decided to sell his own house before buying one in Scotland, but his wife was horrified when she saw her temporary home. 'I couldn't live up here. It's so lonely and isolated!' she declared.

Truth to tell, it's not that remote. There are over a dozen

families living within a mile and a half, and the nearest town is less than three miles away, but the way the house is situated you can't see anyone else.

Living in the country must seem very strange to those used to living in towns or cities. Despite the distance between each other, our rural community is very close knit and most of us know every move our neighbours make. Not for us, as happens in a city, the shame of living next door to someone who has been dead for weeks and never missed. Still, it is true that it is easier to live in the country when you're used to it, and that poses big problems for young farmers.

Once there were plenty of braw lasses for them to court. Many happy love matches began at a dance in the village hall or at a meeting of the young farmers club. Nowadays fewer young folk live in the country. Our lasses go to college or work in town and often marry out of the district. So, our young lads suffer a double handicap: fewer country girls to court and the risk of marrying a townie who, try as she might, simply can't adjust to life on the farm.

That's one of the reasons for the rising divorce rate on farms. Farmers work very long hours and often young mothers are left alone coping with young children. Isolation and loneliness can lead to stresses the marriage can't cope with.

The personal columns of farming papers are full of ads from dating agencies as well as lonely-hearts columns. So, if you are tempted to reply to one, remember that you can marry in haste, but in the countryside you will have plenty of time on your hands to repent at leisure!

2

Bringing home the bacon

Most farms had a sow or two at one time. Kept in a wee, stone-built pigsty, they would feed on the whey from cheese-making as well as on any household scraps. As the saying goes, 'waste not, want not'. As boars are expensive to feed, it was hard to justify keeping one for a couple of sows, so, every so often a young lad would take the sows that were in heat to be mated at a nearby farm.

Modern systems are much more intensive and every year each sow produces twenty-three or twenty-four piglets. In the old days of my grandfather there would only be two litters a year, each with eight or ten healthy piglets. Weaned at about eight weeks, the piglets, or weaners as we call them, were often sold to small-holders or shepherds to be fattened for home use.

Snag with fattening pigs in days gone by was how to store the meat after the animal was slaughtered. The best cuts were steeped in brine and hung up from hooks on the kitchen ceiling to cure as ham. Home-cured ham is delicious compared with most of the modern tasteless stuff that froths in the frying pan.

Without fridges and freezers, the rest had to be eaten fairly quickly. Grandpa's generation solved that problem with communal pig killings where neighbours gathered to help butcher the pig.

Off-cuts and trimmings were made into pies and sausages. Blood was collected and made into black puddings. Even the head was boiled to make potted head. Only the tail and squeal were wasted at a pig killing! Afterwards, the goodies were shared out so they could be eaten fresh.

Gradually the practice died out, and with it the skills of slaughtering and butchering pigs at home. Today you rarely see pigs on farms and, when you do, they are in large numbers on specialist pig farms. The few herds of pigs left in Scotland tend to be found in the arable areas of the east and north-east.

The potential for substantial profits encouraged big business to become involved in intensive pig production. Helped by the demand for cheap, mass-produced food they designed factory systems that took over from traditional farmers. Then things changed as housewives became concerned about the quality of their food and the way it's produced.

In times of surplus the government introduced European Union-inspired legislation outlawing practices like tethering sows and confining them in stalls. That started a trend to run pigs on free-range systems rather than keeping them in expensive concrete palaces. All over Britain outdoor pig units sprang up and fields with rows of pig arcs became common. Premiums paid for free-range pork encouraged the rapid growth in such enterprises. Pigs once again wallowed in muck, rooted about or had a fight to establish the pecking order.

Mind you, feeding outdoor pigs on a cold, wet winter's day is not as pleasant for the farmer as in modern buildings, but then there are always snags. One of the biggest problems with keeping pigs is that prices vary so much. They used to have a regular cycle where a couple of good years were followed by a few years of losses.

That cycle of boom or bust became so predictable that some producers learned to increase their breeding herds when profits were at their lowest to take advantage of the expected upturn. Beef farmers have to wait nearly three years for a heifer to produce its first calf and another eighteen months to two years, or more, to fatten that calf. Gilts, or maiden sows, on the other hand can give birth, or farrow, at eleven months and their piglets can grow to pork weight in three months.

The old, predictable price cycles have all changed now and prices seem to be constantly depressed as a result of imports. One of the biggest problems is that those imports come from some countries where 90 per cent of the farms would not pass British Assurance Standards on animal welfare or food hygiene. It's almost impossible to get a level playing field where imports are concerned. Many British farmers have learned in recent years that their pigs don't bring home the bacon!

Anyway, my wife pestered me for years to get a couple of pigs, so I eventually agreed and bought three 8-week-old weaners for her wedding anniversary one July. Ordinary men would simply have given their wife a bunch of flowers but real men like me give pigs. Truth to tell, I was just as keen to have them as she was. Of all the farm animals I have ever worked with, pigs are my favourite. Not only are they clean but also very intelligent and full of character. It isn't easy to find good pork nowadays, so the plan was to fatten them for our own use. Three pigs are too much for one family, so a couple of friends were to share in the feast.

Although I was sharing, unlike the old days the pigs were taken to an abattoir to be humanely butchered before being delivered to my local butcher to be professionally cut up.

Those three weaner gilts had been runts from their respective litters, but my wife's tender loving care soon had them thriving. Their diet of pig meal was spun out with household scraps and stale bread set aside by local shopkeepers. Although they were ready for slaughter in early December my wife postponed the fateful day, until they became so massive that they had to be dealt with. So one February morning I had a war of wits with those obese pigs.

Pigs can be difficult to drive and have a mind of their own. Unlike cattle or sheep, which are easily herded home, pigs love to charge straight at you and bowl you over. I suppose that's where the phrase 'pig headed' comes from. In these situations, I use a wooden board about a metre-square that looks a bit like a baker's tray. It forms a solid barrier and prevents them seeing to the side and that helps drive them forward. But dealing with pet pigs is never easy. After seven months of friendly chat and having their backs regularly scratched, they knew no fear. End result was that they playfully came towards me, rather than be driven where I wanted them to go.

Experience has taught me not to feed such pigs the night before moving them. When it was time to load them onto my livestock trailer I got my wife to lead them on with a bucket of feed. Hungry, greedy pigs always fall for that one.

All went well until the biggest pig took fright at the shiny, aluminium trailer and bolted. Several times we nearly coaxed her back with her mates, but she always turned tail at the last minute.

Bitter experience has taught me that it's futile to wrestle with monster pigs, so I used an old trick I learned in my student days. I soaked some pig meal until it became a mash that stuck to the side of a bucket. While she was eating out of that bucket, I slipped the handle over her head and behind her ears. That way she couldn't see what was happening.

Still chomping at her mash and blissfully unaware of her surroundings, I gently nudged her up the ramp and onto the trailer. I eventually drove off leaving a tearful wife behind to phone our friends to prepare for their share of the pork.

Over the years pigs have had a bad press. They have been accused of being greedy, lazy and fat so often that phrases like 'greedy pig' are part of everyday language.

Truth to tell, they can be a greedy lot. As a student I worked on a pig farm in Norfolk where I dreaded feeding time. There were five hundred sows running free-range outdoors in batches of about twenty. They were fed compressed lumps of feed, called rolls, on the ground. Idea was to open a bag containing twenty-five kilos, put it on your shoulder, then open the gate and sprint across the paddock sprinkling out a line of rolls. Easier said than done!

Invariably, you slipped or tripped over a sow and would then be trampled under-hoof by twenty greedy sows devouring the contents of your bag.

Even the way they feed sounds greedy. They chomp and slurp at such a rate that everything is gobbled up in a few minutes.

Pigs have a digestive system similar to ours. Ruminants, like cattle and sheep, have to eat large volumes of grass and need to graze for hours on end, so they're constantly on the move. Once a pig has gulped down its meal it can literally laze about till the next feeding time, and pigs are good at lazing about.

They lie flat for hours on end, occasionally grunting with contentment. Best of all, they love to wallow in a muddy hole on a hot day. Greedy appetites and laziness lead to pigs becoming very fat. That was important in the days when people liked to eat fatty, streaky bacon, well-sizzled in its own grease. Fashions change and nowadays most people prefer lean pork and bacon grilled or cooked in a fat-free, non-stick pan. As a result, farmers have bred leaner strains of pigs and fed them healthier diets to stop them getting too fat.

About six years ago, researchers at the University of Wales reckoned that modern diets and breeding programmes were producing pigs that developed symptoms similar to anorexia nervosa. While it only affected a few pigs, scientists reckoned the new disease was on the increase. One bonus of their research was that it helped them to understand human eating disorders. As I said, pigs have a similar digestive system to humans and are roughly the same body weight. So those thin pigs were used to test new drugs for treating anorexia.

Could it be that porkers are due some better press and we will start using phrases like 'slim as a pig'?

Collie dogs are so intelligent they aren't classed as farm animals.

They are part of the management team and are never far from the farmer. The honour of being the most intelligent farm animal belongs to inquisitive pigs. They're forever investigating with their snouts and looking for something to sniff, chew and destroy.

Pigs are like children and are easily bored, and bored pigs can become very bad pigs! They chew the tails and ears of smaller pigs and that can lead to outbreaks of cannibalism. Such incidents often happen when different batches of pigs are mixed for the first time. To avoid fighting, I used to throw a dozen or so empty paper-feed-bags into the pen. That led to a massive scrum as the pigs set about shredding the bags. When it was all over those sweaty pigs all smelled the same and couldn't tell friend from foe.

Pig boredom is at its worst in modern concrete piggeries. Some farmers bed their pens with straw to give the pigs something to root among and chew. Bedding pigs on straw is the best solution, but is impractical for about three-quarters of Britain's pigs. Straw can clog or block muck-handling systems. That's why the Brussels bureaucrats intervened and introduced new laws in 2003 that made it compulsory for pigs to have toys – or 'manipulative material', as they say in Eurospeak – to play with. Failure to comply can lead to hefty fines. So the race is on to design new and better toys for our pigs.

Scientists designed a 'food ball' about fifteen years ago. It was a hard plastic ball with feed in the centre that trickled out as the pigs played with it. Snag was that it was too expensive.

Practical farmers give their pigs hard plastic, indestructible balls or blocks of wood to play with. Others hang carrots above the pens. Chains and lengths of hard plastic are also good hanging toys for them to sniff, rattle, bite and chew.

With roughly three million pigs in Britain there is a lucrative market for those who can design good piggy toys or games. So, will we see an overworked Santa having to deliver extra toys and games to pig farms? Next time you tell the children to tidy up their toys don't tell them their room looks like a pigsty as that will be much closer to the truth than you dare think.

I remember a television news report where an English judge described pigs belonging to a Norfolk farmer as 'neighbours from hell'. The pigs were free-range and had trampled their field into mud in a spell of wet weather. That mud had been washed onto the road and had made a mess.

As with all hungry pigs, there was a lot of noise at feeding time and all the usual smells associated with farm animals. The biggest complaint was that the pigs had mastered the art of escaping from their field. Once free, they rooted up tasty daffodil bulbs from roadside verges and neighbouring gardens as well as leaving droppings on lawns and driveways. No wonder the folk around those parts reckoned those pigs were neighbours from hell.

The judge ordered the farmer to control his pigs properly or face hefty fines, but that's easier said than done.

I remember from my student days that keeping pigs where they were meant to be was a big problem. As well as having strong fences made of wire pig-netting, we also used electric fences. The idea was to run an electrified wire near the bottom of the fence, about nine inches above the ground. As pigs approached the fence, looking for ways to escape, their moist, sniffing snouts touched the wire and gave them an electric shock. After a few nasty frights they learned to keep well clear of the fence, but some learned to charge at that wire with their hairy foreheads. That way they didn't get nearly as big a shock as through their snouts and the impact caused the electric wire to short-circuit against the permanent wire-fence. Then it was simply a matter of rooting up the fence and wriggling free.

One young pig regularly escaped to rummage through dustbins in the courtyard of a nearby stately home. We eventually had her slaughtered and shared the pork amongst ourselves. She may well have been a lot of hassle but we all agreed she was one of the tastiest pigs we had ever eaten.

Like most farmers I love my animals and care for them to the best of my ability, but farming is a business and I am in the business of producing meat. It's important not to get sentimental because ultimately my animals will be slaughtered for butcher meat. Finding dead animals hurts the farmer's pocket more than his feelings because we are used to dealing with death. It's a sad fact of life that if you keep farm animals you have to expect dead ones from time to time.

Putting animals down is different, because as often as not you have tried to nurse them back to health. There's something about caring for a sick animal that brings out the sentimentalist in me. Animals are blessed compared with humans as they don't understand death, but they can certainly feel pain. I never allow an animal to suffer pain once I realise that there is little hope of curing them.

Culls to contain the spread of contagious diseases like foot-and-mouth or classical swine fever are horrendous events that depress most farmers. The last outbreak of swine fever was in 2000 and started on a Norfolk breeding unit that supplied infected pigs to a Suffolk farm as well as another in Essex. Ministry vets reacted swiftly to that outbreak – which was the first in the United Kingdom for fourteen years – by imposing restrictions on the movement of pigs on the infected farms as well as other pig farms in the surrounding areas and any that may have bought pigs from those infected farms. About twelve thousand pigs had to be slaughtered to eradicate the disease. Their carcasses were transported in sealed containers to rendering plants for incineration.

Classical swine fever (CSF) is caused by a virus and is one of eight notifiable diseases of pigs in Britain. The first outbreak occurred in America in the early 1830s, but it has since spread to most pig-producing countries. While CSF is a deadly and incurable disease of pigs it is absolutely harmless to humans and all other animals. Infected pigs run a fever, develop brick-red-to-purple patches on their skin, have laboured breathing, running eyes and severe diarrhoea, lose their co-ordination and ultimately die.

Even in its mildest form the disease can devastate a pig herd, and that is why it has to be ruthlessly eradicated.

Despite being contained by all pig-producing countries, CSF keeps reappearing. Many believe that at the beginning of every outbreak there's a piece of old pig meat involved somewhere along the line. The virus can survive for at least ten months in air-cured sausages, salamis and hams and for five years in frozen meat. Often an infected pig slips through the net at the start of an outbreak, allowing pig-meat products carrying the virus to be exported to other countries. Most farmers know the dangers of allowing their pigs to come into contact with ham or salami sandwiches, but few members of the public do.

Pigs that run free-range and outdoors are at risk: from titbits dropped in their field by the birds that scavenge in picnic sites and lay-bys. As with foot-and-mouth, an outbreak of CSF can lead to a lot of misery.

3

Tied to a cow's tail

Before the development of railways, cows were kept in towns and cities to provide fresh milk. It was difficult to transport milk long distances over the rough roads in those days and it literally churned into butter. Some farmers walked their cows into the towns and milked them at the back of shops that sold milk fresh to customers. Other more enterprising shopkeepers kept a few cows of their own in byres at the back of their shops.

As recently as 1927, Edinburgh still had 97 dairies with 3,750 cows. Byres Road in Glasgow is so named as a result of the dairy cows that were kept and milked there. There was also a 1,000-strong cow herd situated at Port Dundas in Glasgow, on the banks of the Forth and Clyde canal. On a square with Bath Street in the south, Sauchiehall Street in the north, and West Nile Street in the west was the 'Grand Byre' that held 250 cows. When the streets were quiet during the night, hundreds of horses and carts brought grass, hay, kale, turnips, oats and the like into the town byres to feed thousands of cows. Similarly, those carts returned to the surrounding farms loaded with smelly dung from those well-fed cows. With the advent of smooth railways, tarred roads and refrigeration the town byres fell into disuse.

Like most upland farms in our area there used to be a small herd of about forty hardy, little, brown-and-white Ayrshire cows on my farm. In those days cows were kept in stalls in the byre during the winter. Tied by the neck with chains and individually fed on a diet of hay, chopped turnips and concentrated feed, they gave rich, creamy milk. Cows didn't give as much milk then because the feed was often of poor quality. Mouldy hay and inferior concentrate-mixes were partly to blame.

The system was hard work in those days. Maids from the farmhouse helped the farmer and his men to hand-milk the cows. Working by the light of paraffin lamps they would have the morning milking done before sunrise. Then, while the maids cleaned and polished the lamps, the muck was barrowed out of the byre to the midden. Hard and skilled work, as anyone who has tried pushing a loaded barrow up narrow, slippery midden planks will testify.

Because isolated farms like mine were too far from large towns, and there were no bulk refrigerators, the milk was made into cheese on the farm. With the advent of the Milk Marketing Board in 1933 – with its system of collecting milk and guaranteed payment – cheese-making stopped on most farms. The Milk Marketing Boards were formed because dairy farmers were getting a raw deal at the time. Milk is perishable and dairies played on that fact. Dairy farmers were forced to take a low price or pour their milk down the drain. Unlike grain or livestock that can be kept till prices improve, milk soon goes off.

Eventually, by the 1960s, most upland dairy herds were sold off and replaced with beef cows. Farmers were no longer prepared to slave away in old-fashioned byres and building expensive new cattle sheds for a small dairy herd was hard to justify.

Those who stayed in dairying still had to work hard and many still milked in byres. That involved moving the milking machine round the byre from cow to cow. The milk was then cooled, put into churns and taken to the end of the farm road every morning to await collection.

Dairying used to be a good way to get started in farming. The monthly milk cheque allowed many a cash-strapped young man to work his way up the farming ladder, but the industry changed to such an extent that a young man starting to milk 100 cows as a tenant needed at least £500,000 behind him.

In 1933 there were 10,000 Scottish dairy farmers, each with an average of 34 cows. Today there are about 1,400 dairy farmers left in Scotland but the average herd has more than 100 cows.

When I was a young lad, my father kept a couple of house cows that were milked by hand to provide the family with fresh milk, but it's more bother than it's worth these days.

No matter how busy you are on the farm, someone has to spend at least half an hour, morning and night, milking. If you're away for the day, or for a short break, alternative arrangements have to be put in place. Cantankerous house cows have a habit of being at the back of the field at milking time. After walking her all the way to the byre she will invariably wait till she's in the yard before making a mess of the clean cobbles. Once tied in her stall, her udder has to be cleaned before milking. Hand-milking is a laborious job that takes between five and ten minutes, depending on how much milk the cow is giving. It also develops powerful hands and wrists.

Often the bucket of milk was left on the byre window ledge while the cow was put back in her field. Several times I returned to find the milk spilt on the byre floor as a result of thirsty farm cats jumping up. That led to rows with my father who insisted the bucket was brought into the house immediately after the cow was milked.

After milking, the milk was strained through a special sieve to remove any hairs or dirt. Finally the bucket and sieve had to be thoroughly washed and then the dung swept from the yard or byre.

To be sure of a constant supply, most farms kept two house cows that calved at different times of the year. That led to surplus milk being produced when two lactations overlapped. Although some was used to rear calves, there was often waste that was fed to the dogs.

With so much hassle involved, the milkman is a far easier system.

He leaves milk at the end of the farm road. If I want extra, I order it and when I am away I cancel it.

As it has been pasteurised it is perfectly safe to drink. I doubt if I could drink raw milk today without getting a tummy upset from the bugs that I had once built up immunity against. It's only when you've milked a cow yourself that you appreciate what good value milk really is!

As a student I worked on a dairy farm for a few months. Fortunately, it was in the summer, which is the best time of year. That's when the cows are outside, chewing their way across a field, turning grass into milk and coming inside twice a day for milking.

I loved that early-morning walk through the grass wet with dew and the cows getting to their feet and starting to move towards the gate. A grand time to be up and about, but not so much fun when the days get shorter.

Milking twice a day, seven days a week, fifty-two weeks a year, doesn't appeal to everybody. It's called being tied to a cow's tail. I, for one, enjoyed my few months, but wasn't sorry to return to beef and sheep farming. Rising early to milk cows doesn't fit well with a student's social life and some weekends I was hardly in bed before I had to get up again.

Modern dairy farming is high-tech and, as a result, milk yields have virtually doubled since the war. The average Scottish cow gives well over six thousand litres of milk every year. New breeding techniques allow top bulls to be identified and used widely through artificial insemination (AI) – or the bull with the bowler hat as farmers jokingly call it. However, although better breeding practices have been an important factor, the main areas of improvement have come in management and nutrition.

Instead of old-fashioned byres, today's dairy cow is kept in the comparative luxury of a loose-housing system. Instead of being individually chained to a stall, they can wander about at will. Maybe they'll help themselves to silage at the self-feed barrier, or go for a leisurely drink of water. Others may rub their itchy backs against brushes specially designed for that purpose. Instead of lying on a cold concrete floor, modern dairy cows are more likely to lie on rubber-cushion mats that are as comfy as a mattress. You see, the more comfortable she is the more milk she gives.

In a modern cow shed every corner is kept spotlessly clean by removing muck with scrapers on chains that work automatically. Milk from a cow is warm and has to be chilled and stored in big, refrigerated, stainless-steel tanks. Some farms use heat exchangers to take the heat out of the milk and save on electricity bills. The resulting warm water is then piped to the water troughs. Taking the chill off the water encourages cows to drink more and that leads to higher milk yields.

Some progressive farmers have invested heavily in robots that milk the cows for them. Cows go to the robot when they want to be milked and it does the rest. It means the animals can be milked more frequently and that encourages higher yields. Cows that don't have to carry gallons of milk in their udder for up to twelve hours are more comfortable. Better still, the farmer can have a lie in.

Modern dairy herds in the United Kingdom can be in excess of two hundred cows and herds of three or four hundred aren't uncommon. The problem with big herds of dairy cows is that it takes so long to walk them to and from their fields to be milked twice a day. A herd of 250 cows can stretch out for more than half a mile. Big herds also trample their fields into mud during a wet spell. Another problem with grazing fields is that the quality of grass varies according to the time of year.

Some of the larger dairy farmers have found it is more convenient to keep their cows indoors all year round and make all their grass into silage. The idea is to feed the cows a balanced ration that includes silage, but doesn't vary in quality throughout the year. Better still, you don't waste time walking cows long distances. It sounds novel but, as I explained earlier, our forefathers developed that system for cities two centuries ago. There's seldom anything new under the sun!

Genetic progress has also helped to increase yields. Farmers can select semen for artificial insemination from bulls that breed docile offspring. When I was a student I well remember how difficult it was to get freshly calved heifers to come into the parlour to be milked for the first time. Often as not they would repeatedly kick the milking machine off their udder and it would take several days of patient handling to get them to settle down to the routine of being milked. Nowadays that's rarely a problem because generations of dairy cows have been bred for docility and most modern heifers meekly stand to be milked for the first time.

It's also important to have docile animals because fighting and bullying leads to stress and poor performance. Mixing strange animals with a bunch that have been together for some time often leads to fighting. Hens peck each other, pulling out feathers until their stranger victim becomes a bedraggled, bloody mess. It's the same with pigs. All hell breaks loose if you put strangers into a pen.

Cattle can be difficult to mix. Bullocks fight each other with the risk of serious injury. It's almost impossible to introduce a freshly calved heifer into the milking herd without older cows bullying her. That can lead to heifers being pushed upside down into feed troughs as well as suffering injured legs or displaced stomachs. When they go to the trough to feed they get bullied again, so they don't eat and timidly stand to the side. Even lying down is a problem as bigger, older cows bully them out of their cubicles, forcing them to lie down in the dirt.

When heifers are introduced to the milking herd it's important that they settle down straightaway so they're milking well within a few days. Some farmers try to overcome the problem by introducing heifers at night. That often fails because the lights aren't off for long enough and heifers haven't enough time to settle before cows can see there are strangers in their shed.

The Americans may just have found the solution we have been looking for. It's simply a matter of pouring vinegar over the backs of heifers before introducing them to the herd. According to the Yanks the cows just walk away because they can't stand the smell of vinegar and that leaves the heifers free either to choose a cubicle to lie in or go to the feed trough and eat when they want. As is often the case in farming, it's the simplest ideas that work best.

Whether traditional or modern methods are used in dairying, one thing's for sure: milk is as good for you as it has always been. Until the seventies, children were legally entitled to free milk at school. As a boy, I remember how horrible that milk tasted after sitting in crates in the sun on a hot summer's day. Then there was wintertime when the milk froze in the bottle! Despite that, most of the time I enjoyed my wee bottle of milk at playtime. Calcium-rich milk is an important part of any child's diet. For one thing, it helps to develop strong bones. Later in life it helps limit bone loss or osteoporosis, as it's called. Indeed, the NHS estimates the cost of osteoporosis – which affects one in three women and one in twelve men – at over £1 billion per year.

That small bottle of milk at playtime was often the first food a youngster took in the morning, as many went to school without breakfast. Biggest worry now that free school milk has been abolished is the well-publicised potential for children to develop bad diets. They need little encouragement to buy chilled, fizzy drinks; manufacturers know that only too well and make sure most schools and canteens are well supplied. In the absence of an alternative such as chilled milk, it isn't hard to imagine our children drinking more fizzy concoctions with their chips and burgers. That will also lead to poorer teeth and more work for dentists.

Once children got into the habit, as I did, many preferred a glass of milk with their lunch. Many of us now fear that future generations will lose the healthy, milk-drinking habit.

4

Bunny is just a menace

All over Scotland rabbits are once again reaching epidemic proportions because myxomatosis is no longer keeping their numbers down. It's a horrible disease that used to be a real killer. When it first appeared in this area in 1955 there were dead and dying rabbits everywhere and millions died. In the final stages, their eyes swell up and they become blind. A truly pitiful sight!

Rabbits used to be common on farmhouse menus, but all those dying rabbits put me off them and I haven't eaten one since. During the 1960s, rabbits became a rare sight but, over the years, they've become resistant to the disease and many now shake it off as we would a dose of the sniffles.

Mind you, the weather has also played its part in allowing rabbit numbers to build up. A series of mild winters helped, as well as a run of wet summers. Myxomatosis is spread by rabbit fleas, which thrive in hot, dry weather. Another problem is that rabbits have changed their habits. Instead of crowding together in warrens, many now live healthily above ground making it harder for the fleas to spread from rabbit to rabbit.

But the biggest problem with rabbits is . . . they breed like rabbits!

Their main breeding season is from January to July. Although does can become pregnant from January onwards, bucks aren't much interested after July. There are no hard and fast rules and some rabbits can breed all the year round and they are extremely fertile.

A doe can breed in her first year before she's fully grown and mates again within a few hours of giving birth. That means that each litter, of about six or eight young, is separated only by the length of pregnancy, which is about a month.

There is another phenomenon peculiar to rabbits and hares. When litters in normal growth within the doe die before birth, say due to harsh weather, instead of being stillborn they are completely reabsorbed into the tissue of the mother. This process only takes a couple of days, after which the doe will act as if the litter has been properly born. So, when absorption is complete, she conceives a second time and soon has a fine litter.

No wonder numbers have recovered so dramatically. Back in the 1950s, when they were virtually wiped out with myxomatosis, farmers were so glad to be rid of them that they set up rabbit-catching societies so there would be no chance of them returning. A group of farmers got together and employed part-time rabbit trappers. Most were retired gamekeepers, shepherds and the like who, for a few shillings an acre, gassed, ferreted, snared or shot rabbits.

Yes, it's sad but true, that controlling bunnies means killing them. Such schemes fell by the wayside when farming profits became tight. Anyway, the experts reasoned that when rabbit numbers increased, myxomatosis would kill them off again.

Predators like buzzards, foxes and stoats are enjoying a feast, but they are having little impact on overall numbers. At the end of the day, only man can control the rabbit population.

Snag is, even if you could find a skilled rabbit trapper, the cost of hiring him is prohibitive. Worse, rabbits are almost worthless nowadays. At one time they were worth a lot of money. Many are the tales I have heard of big farms during the depression of the 1930s hiring rabbit trappers. After skinning the animals, they were sent by rail to be sold in the cities. Even their skins had a value. I knew one farmer locally who started out life as a trapper and made enough money to start farming.

Apart from snaring, ferreting and shooting, some farmers used to walk the fields at night with long nets. A team handling the net would encircle and trap the grazing bunnies. Many a luckless unemployed man kept his family on rabbits during the hungry thirties. In return for being allowed to snare the animals for the pot, he would help the farmer in the hay or harvest fields.

Apart from the odd sportsman looking for a day's shooting or ferreting, nobody seems interested in the humble bunny today. Farmers haven't the time either to gas rabbit burrows, or to check snares daily. One farmer I know came up with the clever idea of running an electrified wire round his fields, set about six inches above the ground. The high-voltage shock kept the rabbits away, but it was a time-consuming job putting up the wire and cutting weeds back so that it didn't earth.

Apart from mating, bobtail's favourite pastime is eating. Wide strips of crops next to woodlands or railway embankments are nibbled bare. They may look cute but rabbits are very destructive. It's their method of grazing rather than the amount they eat that causes so much damage. They nibble the grass so closely that it doesn't grow again and is soon replaced with mosses that typify a rabbit-infested pasture. It's reckoned that ten rabbits eat as much grass as one ewe. So a farm with a modest infestation of several thousand could have kept another three hundred ewes. Once again it's not just the grass they eat that is the problem. Rabbits graze out the best grass, encouraging poorer types that sheep and cattle don't like, and their droppings and urine contaminate even more pasture than they consume.

Not only do they eat crops, but they also stunt the growth of large areas leading to lower quality, unevenly ripened grain. To give some idea of the damage rabbits can cause, one Perthshire farmer reckoned they were costing him £70,000 a year.

Young trees become deformed after the buds are nibbled and are often killed in frosty weather when rabbits strip the bark. That is why we have to protect newly-planted hedges and woods with expensive netting that has to be dug into the ground to prevent them burrowing underneath.

Railway embankments are favourite haunts as they provide easy tunnelling and dry warrens. That has led to a real risk of railway lines being undermined by thousands of burrowing bunnies. Their burrowing activity also poses a big threat to archaeological sites by mixing the different layers of soil and making it harder to unravel the history.

As I said, the old methods of control are of little avail. One farmer told me once that he and a friend were shooting three hundred a night without any discernible effect. Fortunately, a group of Perthshire farmers developed a new trapping system that is both humane and efficient. The idea is to fence-off woodland and rough banks, where the warrens are hidden, with rabbit-proof netting. The rabbits can only get into the fields through special holes in the fence, where the traps are sited. Those traps are wooden boxes about two metres by one metre by a metre deep, dug into the ground. They work on the principle of a flap on a pivot. The flap is secured until the rabbits are used to running over it. Then once a week the flap is released so they fall into the box. There they lie, contentedly in the company of each other, as if in a dark warren, until they can be humanely destroyed.

It's simple and effective. I know a farm where they caught 22,000 in two years and that makes a lot of rabbit pie.

5

The pecking order

You should never count your chickens before they've hatched. I learned that lesson the hard way. As a young lad my first enterprise involved poultry. I figured that the few bob saved from my hard-earned pocket money would be wisely invested in a few bantam hens. That's how I became a ten-year-old poultry breeder.

It was a serious business cutting turf, laying the sods upside down in old tea-chests, forming a nest and then fitting doors. Once the nests were prepared, broody hens – or clucking hens, or cluckers, as we call them – were set on their clutches and fed and watered daily. Small amounts of water were sprinkled regularly on the turf to keep it moist. That way the shells didn't become brittle and the chicks could hatch easily.

As the days dragged by my childlike mind calculated the profit from the cockerels I would sell to mother and the increased egg sales from the next crop of pullets. Sadly it never worked out that way. There were always infertile eggs and chicks that died through trampling, drowning or catching a chill.

As I said, I learned at an early age that farming is very unpredictable. So why is it that so many so-called experts still indulge in the practice? No sooner are our crops planted than there are predictions of a bumper harvest. The experts never take into account the real possibility of the crops failing because of a cold wet summer, or being flattened by a wet, stormy autumn. I suspect a lot of those predictions are made to talk down the grain market and encourage farmers to sell-on forward contracts at lower prices.

Anyway, my wife has always kept hens. She started with about a dozen free-range ones that ran about the yard and fields scratching for worms, insects and seeds. In the morning she fed them a wet mash that sometimes had household scraps added, such as stale bread or boiled-potato peelings. In return they laid enough eggs to feed our hungry family. And what eggs they were! Big brown eggs fresh from the nest boxes, with their huge, dark-orange yolks that put shop eggs to shame. Scientists will tell you they have the same nutritional value as eggs from battery cages, and that may be so, but my wife swore by them for baking, and reckoned her sponge cakes rose better with a nice, creamy, yellow colour. A quick glance at my portly waistline at the time proved there wasn't a lot wrong with the baking in our farmhouse.

Eggs come in all shapes and sizes: normal, round, elongated or conical like a peewit's. They can also be massive with double-yolks, or as small as a cherry. Sometimes their shells are so thin that they crack on boiling. I have seen them without a shell, just held together by a thin, leathery membrane. Occasionally, you get them with a wrinkled shell. It's all to do with the hen's age, diet, health and the precise stage in the laying cycle.

Pullets start by laying small eggs with tiny, pale-yellow yolks. Best chance of getting eggs with double yolks is in the first month after they start to lay. Eventually they settle down to produce big eggs, with rich, deep-yellow yolks. Lack of calcium in the diet leads to thin, brittle shells, but is easily cured by feeding ground shells from the seaside. Hens suffering from respiratory infections often lay those eggs with wrinkled shells, although the contents are perfectly okay.

Every year a hen moults to grow new feathers and stops laying eggs for a while. You are most likely to get extra-small or extra-large eggs in the weeks prior to the moult.

The eggs you buy in the shops are produced under strict hygienic conditions. Most are now laid in 'roll-away' nest boxes. After the egg is laid, it gently rolls out of reach of the hen to be collected. That keeps it clean and away from hungry hens. Oddly enough, some hens also enjoy an egg for breakfast. Apart from the loss, it's not hygienic to have egg white, yolk and nest material sticking to other eggs.

All shop eggs are thoroughly examined before being packed. As they are graded according to size, they are passed through strong beams of light that show up any cracks. Mother Nature has made the egg virtually impenetrable to bacteria. It has to be to protect the developing chick. Cracked eggs can become contaminated and are rejected.

Mind you, there's not a lot wrong with a fresh, cracked egg when it is properly cooked. My wife always sold the good eggs and kept the cracked ones for baking. It's all a far cry from grandmother's day when hens laid in hidden nests all over the farm. A number may have lain a long time before being found. So granny gave them the water-bowl test. Eggs that floated were off! As a lad I threw such eggs at the dyke and ran laughing from the stink.

Our hens then, as now, were a modern hybrid that we have always bought from a specialist breeder as point-of-lay pullets. Because a hen will stop laying eggs when she's clucking, the breeders have bred that trait out of modern birds. The ideal modern hen is supposed to lay at least 250 eggs a year, and hasn't time to brood.

Despite that, after a few months running free on the farm, some of them forgot that they were genetically engineered superhens and instinctively went back to their roots and occasionally one would start clucking. When that happened we used to catch the clucker and put her in a nest made in an upturned tea chest. Then we slipped a couple of china eggs under her and let her sit on them for about a fortnight, after which we would buy a dozen day-old chicks from a specialist hatchery nearby.

Those chicks that were specially bred for the table were quietly slipped under the hen and the china eggs removed, fooling her into thinking she had hatched them. She would then rear and care for those chicks as if they were her own.

Every time she found something nice to eat, she would cluck furiously, bringing all the cheeping chicks at the double. When it turned cold and wet she would gather them under her wings for warmth, and, if a farm cat hungrily watched the young chicks, she would fluff up her feathers, launch into a fierce attack and chase it away.

Once they had grown feathers, we would take those chickens from the hen and fatten them in a separate pen. They always grew into big succulent birds, the like of which I have never tasted since.

I remember once slipping half-a-dozen day old ducklings under a hen. She proudly looked after them until they were about a fortnight old, at which point they discovered the nearby burn. Without hesitation they all dived in for a swim, leaving a worried hen on the bank. I laughed at her clucking away furiously, telling them to get out of the water before they caught a cold, but, of course, they never heeded.

Incubators and artificial-rearing units are all very well, but you can't beat Mother Nature's way of doing it. I remember another incident when we were invited to dinner by a couple of 'white settlers', who lived in a nearby village. As we sat having a drink we couldn't help hearing the noise of a young chick coming from another room.

Our friends explained they had a cat that was in the habit of catching small birds and animals and bringing them into the house alive. Just before we had arrived it had brought home a day-old chick. The little chick was black, and although I couldn't be sure, I told him it was probably a bantam chick. After a general discussion we concluded it had probably been taken from the yard of another white settler who kept hens.

Young chicks need a constant source of warmth or they soon die from hypothermia. After filling a hot-water bottle with warm water, and wrapping it in a tea towel, we placed the wee mite on top to keep it snug. Throughout the meal he chirped away, not realising that few chicks survive a tussle with a cat. After the meal my wife took that chick home and fostered it onto a clucking hen. She gave the mother and chick to our friends who erected a cat-proof pen to keep them in. That wee black chick grew into one of the noisiest bantam cocks I have ever known. It woke everyone in that village at the crack of dawn for years.

As time passed my wife kept more and more hens. At Easter, when the birds tended to lay well, she would sell surplus eggs to friends. Word got round and over the years demand built up until we had a waiting list of customers. At her peak she kept a flock of over two hundred birds in three henhouses that normally supplied eighty dozen eggs a week for sale locally.

As the flock built up we had to buy henhouses to replace the rickety one I had made to accommodate a dozen. The first one we bought held sixty. At just over £1,000 delivered and erected, it was an expensive choice, but what a henhouse! It had plenty of windows to let in light and, more important, it was well ventilated. Henhouses can get very stuffy on a hot summer's night when all the birds are roosting.

The hard work began once we had installed sixty 18-week-old point-of-lay pullets in that state-of-the-art house. Hens have to be trained to use the nest boxes, or they'll lay eggs all over the farm in hidden corners. To do that, we kept them shut in for about three weeks till they were all laying eggs where they should. Another wee trick that helps is to put china eggs in the nests, to give them the idea.

Most pullets are reared on deep litter and have never used perches. So every night for the first couple of weeks some of them have to be lifted off the floor and placed on the perches! Once properly trained, they were let out in the morning after most of the eggs had been laid. They didn't stray far to start with, and came back to the henhouse to lay and roost at dusk. Mind you, there was always one silly so-and-so that refused to go in, but eventually they settled down to a routine. That was when the last important chore of the day took place, shutting them in to keep them safe from foxes.

I remember one occasion when a fox visited the henhouse and killed one of the birds during the night. When my wife went to feed them the next morning there was a trail of brown feathers on the farm road that ended in a half-eaten carcass. Rather than struggle into her waterproofs and face the blast that stormy wet night, my wife took a chance and didn't shut her hens in, but hungry foxes are always on the prowl whatever the weather.

Fortunately, the hen that was killed was one that was bullied and slept on the floor near the door rather than on the perches with the rest. That's one of the snags with free-range hens. From time to time they pick on smaller or weaker birds. I suppose that's where the term 'pecking order' comes from. Anyway, that timid hen had learned to keep itself to itself and sadly was probably the first hen the fox saw when it peeked through the door. More than likely that fox was a half-grown cub learning to fend for itself. Luckily it only killed the one as young foxes can get carried away and needlessly slaughter scores of hens.

We ran the flock in two age groups and culled half every year towards the end of their second laying season. The trick was to buy the replacement point-of-lay pullets in late summer. Buying them at that time ensured a regular supply of eggs through the winter when free-range ones were in short supply.

It all went horribly wrong one time when we took delivery of a bad batch of feed one November. Hens are pernickety feeders and if they are not fed properly will stop laying and moult their feathers. Once in moult, it takes several months before they start laying again. By the time we realised that the feed was dodgy, the damage was done. All three henhouses were littered with feathers and roosting, moulting hens with no intention of laying.

Fortunately, we dealt with reputable feed merchants who acknowledged there was a problem at the mill, which had, in consequence, produced a bad batch of feed. They apologised and offered to compensate us by buying another 120 point-of-lay pullets. At about a fiver each that was fair as those birds had stopped laying at their most profitable stage. At peak production our type of hybrid hen, called the ISA brown, can lay a large brown egg every day. Once moulted, we faced the prospect of feeding them till the following Easter for no income.

At that time the flock guzzled about £200-worth of feed a month, so we could easily have been £700 or £800 out of pocket before we had anything to sell again. By replacing half the flock when we did, we were selling eggs again in January. That helped my wife's finances and put a smile back on the faces of all her disappointed customers.

We eventually overcame the problem of seasonal egg production with artificial light. Breeding cycles for many animals are triggered by the amount of daylight. Sheep, for instance, come into season in autumn in response to the shorter days. That way their lambs are born in the spring when, with a bit of luck, the weather is kinder and grass is starting to grow.

Most birds, on the other hand, mate in the spring when the days are getting longer. As a result, their eggs hatch at the start of summer when there's an abundance of insects to feed their young. Hens are no different and also produce most of their eggs in the spring. To overcome that problem, modern poultry farmers control the amount of artificial light in their laying sheds. As a result, they can keep their egg supply fairly steady throughout the year.

My wife nagged me for years to install electric lighting in her henhouses. Easier said than done!

Our henhouses are about four hundred metres from the farm buildings to discourage the hens straying into the yard and laying eggs in secret nests in the hayshed. They're encouraged to lay in the nest boxes provided in the henhouses. Snag was, how to get electricity out to those henhouses.

One idea I toyed with was to use old tractor batteries and recharge them during the day, but that sounded like another tedious chore in an already busy day. Anyway, tractor batteries aren't light and the thought of lifting them in and out of the Land Rover didn't appeal. So I decided to run an electric cable from the farm buildings on second-hand telegraph poles.

A JCB had to be hired to dig the holes and help place the poles in position. Then an electrician had to be hired to wire up the henhouses and fit an automatic timer. It all added up to a lot of expense that took several years of extra egg production to pay off.

The timer is set to switch the lights on at two in the morning and switch them off again at daybreak. In the dead of winter that fools the hens into thinking it is really spring and keeps them laying.

The system worked reasonably well over the years until a power cut left our hens in darkness one morning and my wife found the floors of the henhouses covered with eggs. Our poor hens had laid their eggs at the usual time but were unable to find their nest boxes.

Our electrician was busy at the time and couldn't come for a few days, so we hung lanterns from the roof and switched them on at five for a couple of mornings. The fault turned out to be in the overhead cable that had been broken internally where it leaves the main farm building. That was a result of a pheasant accidentally flying into it on a shoot day.

Overfed, broiler-reared pheasants keep crashing into my power cables when they are flushed out of the woods. Many of the shooters are poor shots who regularly miss those low-flying birds and hit my cables and telephone lines with shotgun pellets. The incident with the henhouses made a change from being without a telephone for several days waiting for the lines to be repaired.

Apart from such unusual snags, many poultry farmers have

developed free-range hen enterprises on a large-enough scale to be profitable. Free-range eggs are more expensive to produce than those from battery-cage systems, but many housewives are prepared to pay a hefty premium for them. Our hens started to lose money despite those premium prices. Buying hen meal in small quantities is expensive. Bigger businesses can buy in bulk at huge discounts. Replacement pullets that cost us £5.50 each were much dearer than those bought in batches of thousands.

Free-range hens may look happy to the uninitiated, but they have a hard life. They have to cope with wind, rain and snow on our exposed hill farm. Disease was common, thanks to mingling with pheasants, sparrows, starlings, rooks and crows that helped themselves to the hen meal in their troughs. Even those poultry units that keep their hens indoors have to be vigilant to keep disease out.

Hungry sparrows aren't daft and are constantly watching for a chance to eat hen meal. If a henhouse door or window is left open they soon pop in for a quick snack. Those few minutes are all it takes to contaminate feed or infect chickens and hens.

Bullying, feather pecking and even cannibalism can be other problems with free-range birds. The last batch of pullets we had was particularly aggressive. There were also losses as a result of road accidents. Hens have no road sense and often dash in front of passing vehicles.

In the end we decided they weren't worth all the hassle and we now only have about a dozen to keep our own kitchen supplied with delicious, fresh eggs.

6

Bubblyjocks

At one time you would see free-range poultry on most farms, but they gradually disappeared as intensive production methods were developed. Though economics were the main reason for the decline, turkeys had an additional problem: modern strains can't breed naturally. Originally descended from North American wild turkeys, they developed into three main types: the Mammoth Bronze was the largest and most commonly kept; then there was the smaller, but better-fleshed Cambridge Bronze; and finally the white turkey.

Those old-fashioned breeds were difficult birds to rear as hen turkeys are poor mothers. Even when hens and bantams were used as surrogate mums to rear turkey chicks there were problems. They perished from a hundred different causes and you had to get used to the daily disappointment of finding yet another one lying dead. You see, turkeys are in the same league as fickle sheep when it comes to finding new ways to die.

Gradually scientists found ways of rearing turkeys intensively indoors. At the same time they altered the shape of the birds through selective breeding programmes. The end result is those beautifully fleshed, broad-breasted turkeys we see in today's supermarkets.

Poultry breeders have made dramatic progress. Modern fast-growing birds that efficiently convert feed provide housewives with affordable chicken and turkey. Supermarket prices offer delicious value for money. Instead of unaffordable luxury, turkey and chicken are now cheap, healthy, everyday meals. In fact, supermarkets often slash the prices of turkeys lower than pet food to tempt customers through their doors in the run-up to Christmas.

One of the snags of genetic progress is that the breast of a modern male, or stag, turkey is so large he can't mate naturally. Nowadays modern hen turkeys have to be fertilised by artificial insemination. Therefore, although turkey chicks are more plentiful than ever, if you buy modern strains to rear you must slaughter them. Unless you know how to do artificial insemination you can't breed from them.

We always used to have some of the old-fashioned Bronze-type turkeys, three hens and a stag, running free range. Idea was that they would hatch enough chicks to provide us with a treat on special occasions like Christmas and Easter. The first year I had them was a blank as all the eggs turned out to be infertile. That was to be expected, as I had also bought a young turkey stag. The following year, around April, the turkey hens again disappeared as they started to brood their hidden clutches of eggs. After about a month they reappeared without any chicks. I eventually found two of the nests and they were full of infertile eggs.

That was a big disappointment. Not only did it look like I was going to miss out on having a home-reared Christmas treat two years running, but also I didn't have any young breeding-stock replacements, and that was important as it was becoming increasingly difficult to find hardy, free-range turkeys.

There is something about such birds. The gobbling and drumming of a turkey stag, or bubblyjock as we call them, as he rounds up his hens and makes a display is a sight I always enjoy watching. Nevertheless, that Christmas dinner turned out to be the two-year-old turkey stag, and he was replaced by a more passionate bird.

The new stag did his job well and one of the turkeys successfully hatched six chicks from a clutch of ten eggs the following summer. Two of the eggs were infertile, while chicks in the other two failed to hatch. Turkey chicks are notoriously delicate. A shower of rain is enough to give them a fatal chill and they can also succumb to infections like coccidiosis. On top of that, there's always a farm cat lurking nearby on the lookout for a tasty snack.

So once those chicks appeared we transferred them and their hen turkey to the safety of a wee shed. That's where they stayed for a couple of months until they were able to fend for themselves.

Mind you, catching those turkeys wasn't so easy, despite the assistance of my wife and two daughters. After bravely chasing off my two young daughters, mother turkey realised we were also trying to catch her and flew off. So I gathered up the chicks, put them in a box in the shed and left the door open. Sure enough, mother turkey was drawn into the shed by the sound of her cheeping brood.

After being released to grow out naturally two of them were eaten for Christmas dinner. That may have seemed excessive, but there's not a lot of flesh on those natural birds. Free-range, bronze-breasted turkeys look big, but when plucked and ready for the oven they can be disappointing in size. Between four and seven pounds was the usual dressed weight. They are also scrawny to look at as they don't have the massive plump breasts of supermarket birds, but their flavour was always out of this world.

Christmas is the time of year when turkeys and the farmers who rear them get in a flap. Obviously turkeys didn't vote for Christmas, but you would be amazed at the long hours farmers put in preparing them for their customers. As there's only one Christmas Day in the year, the pressure is on to have all those turkeys ready on time. Talk about a flap! In the run-up to Yuletide there's the unpleasant task of plucking them, often late in the evening, to fulfil orders. That brings back memories of itchy, fine feathers up my nose and aching fingers.

It's also difficult having turkeys ready at a variety of weights to match the different orders on the day. After all, there's little point having a shed full of heavy birds when half the customers ordered light ones. Most families want birds weighing about four or five kilos. 'I'll never get that big brute into my oven,' is a disheartening complaint in a year when the birds have grown exceptionally well. On the other hand, hotels often grumble that their birds are too small. No wonder that so many of us have packed it in as a bad job.

You simply can't beat a well-hung, farm-fresh turkey for flavour. Frozen ones from the supermarket may be a lot cheaper, but you only get what you pay for. Don't forget, it's an expensive business rearing turkeys.

There are no hatcheries left in Scotland. As a result, all turkey chicks have to be imported, mostly from France. They aren't cheap and can cost upwards of £4 each. Chicks will survive long journeys without food or water until they are three days old. The secret with young turkey chicks is to get them to eat and drink as soon as they arrive.

Some farmers used to mix treacle in warm water to tempt them to drink before feeding them special crumbs. Others took the easy, but more expensive, option of buying 'off heat' poults (older chicks that are starting to grow and to develop feathers). Turkeys also have growing appetites as they get bigger and they can fairly scoff their way through a lot of expensive feed. Farm-fresh turkeys don't fatten on fresh air.

Another snag with turkeys is that they have 360-degree vision and see more than is good for them. Being highly strung they're easily frightened and will take flight at almost anything out of the ordinary. Frightened turkeys will huddle together or pile on top of each other in a corner and smother weaker ones. Low-flying aircraft, a door banging shut or even dogs barking are all guaranteed to precipitate such incidents. Others collide with walls and feeders and bruise themselves. They often appear to be none the worse for their accident until they're plucked. Turkeys with bruises on their breast don't look appetising and are almost impossible to sell. With quality, free-range birds selling for about £50, it doesn't take too many rejects to hit profits.

Another favourite trick of inquisitive turkeys is to jump up and poke their heads through a wire-mesh fence, but often they get stuck and hang themselves. As sure as frogs are waterproof, turkeys are the most fickle of all farm creatures!

7

Red sky at night

Sunrise and sunset are the two best times of the day on a farm. Unfortunately, sunrise is often a time for getting on with work and I am a great believer that an hour in the morning is worth two in the afternoon. After all, you can fairly fly through your work if you start early in the morning.

Sheep and cattle are waiting to be fed at the troughs, yet later on they will have moved away to graze and are reluctant to come back again. Most ewes give birth to their lambs at daybreak, so if you are out among them at that time in the spring you're more likely to do some good and save lambs. Don't forget that in the darkest months of winter, animals that stay outdoors can't be seen for more than half the day except by torchlight and daybreak is the first chance to check them properly and see that they are all well. Despite being busy, there is always time to observe and enjoy the beauty of sunrise. Truth to tell, there's something about dawn that's invigorating. Even the wildlife feels it, as everywhere birds are singing their dawn chorus.

Solitary and mysterious hares head back to the hill to lie up for the day, after spending the night grazing the sweeter grass in the fields, and everywhere there's a buzz of activity that only happens at dawn. Out on the hill, curlews make their haunting, eerie cries while snipe produce a distinctive drumming sound with their wings. Lapwings acrobatically rise and tumble in the air plaintively calling out 'pee wee', from which they get their nickname.

Plucky peewits are my favourite bird and I always look forward to March when they return from their winter quarters to nest on my farm. They're the bravest little birds I know and woe betide the sheepdog who runs too close to a nest as mother peewit and her mate will swoop and dive till it's chased away. When their chicks are hatched, they'll pretend to have a broken wing to lure hungry predators away from their brood. Then, just as dog or fox think that they're about to catch an easy meal, they fly away.

We used to have over fifty pairs nesting on my farm, but numbers started to decline in the early nineties until 1997 when none nested. Maybe a series of cold, late springs killed off their chicks; changes in my farming practices may also have had an effect.

At one time I grew grain and turnips, until falling profits and a lack of farm workers to help hoe the rows of turnips and harvest the grain persuaded me to stop. Newly-sown fields of spring cereals, or the previous year's turnip field waiting to be ploughed, provided ideal nesting conditions for peewits and allowed them to conceal their camouflaged eggs from scavengers like carrion crows, or corbies as we call them. Those big black birds are a deadly menace at lambing time as they mutilate newborn lambs with their vicious beaks. Their numbers have been steadily increasing and they systematically scavenge the fields in search of eggs that they greedily devour, rarely missing a nest.

At one time, whenever corbies flew near a peewit's nest the male and female peewits took to the air to chase them away. That left the nest exposed so other corbies could eat the eggs. By the late nineties I noticed the peewits on my farm change their tactics as the hens sat tight on their eggs and left their mates to fend off the raiders. That seemed to work and numbers on my farm have steadily risen in recent years. Another important factor was that gamekeepers have been trapping and shooting more corbies so the peewits haven't been under as much pressure recently.

Despite my local success, peewit numbers nationally have halved in the last fifteen years, but they aren't the only birds in crisis. Turtledove numbers are down 80 per cent in the last thirty years and skylarks are back 75 per cent in the same period.

Some farmers reckon that the increase in birds of prey like sparrowhawks has helped kill off skylarks, but once again I have to admit that changing farming practices may have had a bigger influence.

Nobody is really sure of the causes. It may be due to predators, it may be changing farming practices, or climate change and pollution. More than likely it's a combination of all these factors.

There are also natural phenomena that affect wildlife such as adverse seasons, parasites and disease, or even food supplies like voles. Mice-like, short-tailed voles are common throughout Britain and periodically there are plagues of them. Voles have short tails and their heads aren't as pointed as mice. It's one of nature's mysteries, but from time to time their numbers dramatically increase over one, two or three years. Just as mysteriously such plagues suddenly disappear and they become scarce for a while. When they are plentiful, foxes and owls dine well and soon get into fine condition so that they breed successfully, but a collapse in vole numbers is equally catastrophic for those that feed on them.

Winter is always a hungry time for farmland birds as food supplies become scarce. Many of them depend on weed and grain seeds left behind in the stubble fields after harvest. As winter progresses, those food supplies are either eaten or ploughed up.

Large flocks of finches used to be common, but their numbers have declined as farming practices changed. Modern combine harvesters are so efficient they leave hardly any grain in the stubbles and modern farming systems virtually eliminate weeds and their seeds.

At one time one of the main lifelines for birds was turnip fields that were full of weed seeds. Sadly, turnips are a labour-intensive crop and few are now grown in this age of staff shortages. Many farmers recognise the problem and try to help the birds by scattering waste grain onto farm tracks or at the edge of fields.

Others roll out old bales of straw full of weed seeds for the birds to peck at. Mind you, such feeding sites have to be chosen carefully as hungry farm cats are attracted to them and badly sited ones are ideal for ambushing birds as they feed.

Although the rise and fall of certain species are easily understood, lots of once-common birds have been quietly and inexplicably disappearing from the farming scene. Starlings are not as common as they once were and even wee sparrows, or spuggies as we call them, are down by 65 per cent. It's been some years since I heard their chatter or saw them nest in my farm buildings. Yet not so long ago there were so many that they were pests. Spuggie droppings in the byre or barn were always reckoned to be a way of spreading disease.

Another big loss in the uplands has been the dramatic decline in black-grouse numbers. I haven't seen them perform their spectacular displays and mating dance on their lek (the patch of ground the birds return to every year to mate) on my hill for several years. The name lek is believed to have been brought to Scotland by the Vikings and is derived from the Swedish word *leka*, meaning to play. Black grouse are common in Sweden.

The male birds perform ritual displays and the females come to choose their mates. In full display the blackcock (the male of the black grouse) is an impressive bird with his shiny, blue-black plumage glistening in the early-morning sun. Each bird struts in his patch of the lek. A quick rush towards a rival is followed by a little whirring jump, then back to the centre. It's easy to see where some of our formal group dances come from.

I had an elderly friend – who has been dead a long time – who loved to watch the blackcock displays on my lek and predicted their decline. It's the same with their smaller cousins, the red grouse. Like many wild bird populations there are good and bad grouse years. Often as not, the poor years are caused by a spell of cold, wet weather killing off chicks. Sadly there is also a long-term decline in grouse numbers. That's partly due to poorer heather as a result of overgrazing or heather beetle and an increase in predators like hen harriers; but one of the main reasons is the increased numbers of ticks.

Ticks have an unusual life cycle that involves feeding on blood once a year, until they die after laying eggs at three years old. They crawl on to grass leaves and wait for a warm-blooded animal like a deer, hare, sheep, grouse, dog or human to pass by. Then the ticks attach themselves to their victim and gorge themselves on blood for several days, depending on the stage of the life cycle. Once full, they fall back to the ground where they lie for another year and develop to the next stage.

Like mosquitoes, the real problem with ticks is the diseases they transmit, such as louping-ill, tick-borne fever and tick pyaemia in sheep. More worryingly is when they bite shepherds, gamekeepers and hillwalkers, because ticks can also transmit Lyme disease. It can cause chronic arthritis and can also attack the central nervous system and the heart.

Ticks can be a real killer of grouse chicks and are probably one of the main reasons for their decline. Ticks often bite chicks near their eyes, causing the eyelids to puff up making them temporarily blind and an easy target for predators like hen harriers.

Many large estates have removed sheep from their hills to prevent the over-grazing of heather and, as a result, to encourage grouse numbers. That has proved disastrous for grouse in a lot of cases.

You see, shepherds dip their sheep in a solution of pesticides that kill off ticks. Better still, the dip persists in the fleeces for three weeks or more, but once the sheep have gone there's no way of controlling tick numbers and it's very difficult and labour-intensive to reintroduce sheep to the hills.

An old method of calculating the number of ticks was to drag a wool blanket over a measured distance and then count the ticks attached to it. Nowadays those of us who walk the hills know from experience that more ticks are attached to our trousers, so it's easy to imagine how much grouse chicks suffer.

Farmers are now beginning to work with organisations like the Royal Society for the Protection of Birds (RSPB) before the birds we have taken for granted disappear forever. More importantly, farmers are increasingly signing up for grant-aided environmental schemes that maintain and improve habitats for wildlife.

I first became involved with environmental schemes about twenty-five years ago when a naturally regenerating, ancient wood on the steep banks that surround my farm was designated a site of special scientific interest (SSSI). It's about forty hectares of deciduous trees that was traditionally used as a sheltered area where beef cows were over-wintered and fed in accessible clearings. That was all right in the days of plentiful farm staff, but I was keen to develop a farming system without employees and built modern, convenient cattle sheds to in-winter the cattle.

I have always believed that a farmer's job is to look after his animals not to look for them. You would be amazed how much time was wasted looking for sick cattle and sheep that lay low in those woods. Worse, they were invariably lying in a steep part of the wood that was inaccessible to farm tractors, making it impossible to fetch them out so they could be properly nursed in the farm buildings. As a result, I fenced off the woods to keep the cattle and sheep out and to confine them in fields where they were easier to check and look after.

You would be amazed how quickly the natural flora of the wood regenerated once the farm animals were excluded. In the absence of trampling hooves and constant grazing, a whole host of plants – like our native Scottish bluebells, primroses and rare orchids – reappeared. As those plants re-established so did a whole host of insects that were an important food source for birds. Fallen timber – which is a rare feature in commercial forestry – is left to rot and becomes ideal habitat for invertebrates that feed birds like woodpeckers.

I often walk through those woods where the spectacle of death and decay fascinates me. Trees fall and lie upon the ground for years. I can lift up parts of rotting trunks and branches to let them crumble in my hands and fall to the ground as compost.

Once a seed, then a sapling, then a mighty hard tree – now softly turning into earth from which will spring another seedling. How many weary and sick animals have lain down in that wood to return to the ashes that never die?

It didn't take long for Scottish Natural Heritage (SNH) to notice what was happening and designate the area a SSSI. Part of the deal for maintaining the fence and excluding the farm animals from the SSSI was an annual payment that fairly compensated me.

Over the years it has evolved into a very special and beautiful sanctuary, where all kinds of birds, animals, insects and plants can be found. At night, barn and tawny owls silently hunt the edges of the wood in search of voles, whilst badgers come out of their setts to play and feed.

In the morning there is the dawn chorus of birds like song thrushes, woodpeckers drumming and cuckoos. They are interrupted by the excited chattering, yickering and barking of red squirrels, which can easily be mistaken for quarrelsome jays and magpies. Overhead, buzzards, kestrels, sparrowhawks and occasionally peregrine falcons patrol the skies in search of prey. Rabbits scamper near their burrows, whilst shy roe deer silently keep watch over young fawns hidden in the undergrowth.

My landlord rears thousands of pheasants in those woods and most are shot by paying guests. I used to enjoy a day's rough shooting for the pot when I was younger, but must confess that as I have grown older I have turned against the so-called sport of shooting driven pheasants. I now reckon that shooting clay pigeons is a better, and completely harmless, way of testing your marksmanship with a shotgun.

Some pheasants are crafty and learn either to run through the woods or to fly just a few feet above the ground. Such birds are not considered sporting, so nobody bothers to shoot at them. Most fly high and fast till one day their luck runs out and they are shot and killed by a marksman.

There was a cock pheasant that outwitted the shooters for several seasons and during the winter waited patiently every morning at the troughs where I fed my rams, or tups as we call them. He was a real beauty, with magnificent plumage, and he knew it! He arrogantly strutted his stuff and cackled loudly to draw attention to himself. When I sprinkled the feeding into the troughs, he fearlessly competed for his share with the hungry tups.

Despite being bold and arrogant that old bird was no fool. Whenever the gamekeeper's Land Rover appeared with a party of shooters, that cock pheasant slyly hid under the garden shed at the side of the farmhouse. That way, the beaters and their dogs didn't flush him out with the rest and he was spared having to fly the gauntlet of the waiting guns. As soon as the shoot was over he boldly reappeared to strut his stuff.

Odd though it may seem, shooting parties and gamekeepers are in a major way a bird-lover's best friends. Upland farms like mine have little natural food during the winter for game birds to feed on. As a result, the gamekeepers set up hundreds of self-feed hoppers throughout the woods that are regularly filled with a mixture of wheat and protein pellets. That ensures the pheasants and partridges are well fed, but it also provides a regular supply of food for all the other birds that live in my wood. As a result, my farm supports an ever-increasing number of small birds that would otherwise die during a harsh winter. Most prominent among them are thrushes that have made a dramatic recovery on my farm.

Organizations like the RSPB simply don't have the resources to ensure all of Scotland's songbirds are as well fed as mine. It's just a pity more pheasants don't keep their heads down on shoot days!

Another important environmental scheme involves a Land

Management Contract (LMC) that was introduced by the Scottish Parliament in 2005. One of the options in the scheme provides for an annual payment towards the costs of identifying and maintaining continuous paths across farmland. Such routes have to be suitable all year round for walkers and, if the surfaces are appropriate, for cyclists and horse riders as well.

My farm road is about three miles long and runs through the middle of my land. It's a popular walk for locals. Occasionally someone doesn't shut the gates properly, or leaves litter, but most folk are considerate and do no harm. It's grand to see folk enjoying nature and the spectacular views from my farm.

I think LMCs are a great idea and have agreed one to erect signs on my road and maintain it in return for an annual payment of £3,500. That way I am now being fairly rewarded for providing a lot of pleasure to ramblers, cyclists and horse riders. Incredibly, it's reckoned that about thirty-eight million people walk for pleasure at least once a month. It's also estimated that more than seven million of us walk every weekend in the countryside. That's an awful lot of ramblers!

At one time landowners would have discouraged such intrusion with signs like 'Keep Out', 'Private Property' or 'Trespassers Will Be Prosecuted'. The truth of the matter is that there really never was such a thing as trespass in Scottish law. That selfish concept was an English one. Scots law granted a liberty to roam to everyone, provided they respected the landowner's privacy and didn't damage anything.

That all changed early in 2005 when the Scottish Parliament granted everyone the right to roam. It's subtly different from the old liberty to roam and has also had the effect of widely publicising the fact that the Scottish countryside is there to be enjoyed by all. I never understood the selfish attitude of English landowners who kept the public off their land. People have fought and died defending this land, as well as subsidising it in times of peace. So no one should have the right to deny access to considerate folk who respect crops, livestock and privacy and cause no harm or damage. Indeed, it would be absurd for me to accept payments from the public purse to enhance the environment and then not let the public enjoy it.

There is now a good future for wildlife in this country as a result of all the environmental schemes and farmers' changing attitudes. We can regain much of the wildlife we thought we had lost and recreate the countryside of my youth. A wonderful example is the corncrake that recently faced extinction in the United Kingdom, with only a few survivors on some Scottish islands. Corncrake-friendly crofting and farming has doubled their numbers and their future is now much brighter. For my part, I now find that environmental payments have become an important source of income on my farm as well as improving biodiversity.

Scotland has much of Europe's last wildernesses and we have an exciting environment full of life, movement, sound and colour that many of us take for granted. In built-up countries like Holland, city dwellers need horizons that stretch further than the next block of high-rise flats. They need to be at one with nature to find an inner peace.

It's amazing just how important wildlife is and how it adds something to the day. For instance, there is a robin that comes into the cattle shed at daybreak during the winter for some warmth and a bite to eat from the cattle troughs. He loves to fly into the tractor cab when I am feeding the cattle their silage and makes me laugh when I see him sitting on the steering wheel as if he intends to drive that mighty tractor. Unfortunately, he always leaves a wee message on the steering wheel when I chase him away. To avoid having to clean up a small mess, I have to keep the windows and doors of the cab shut when I'm forking-off silage.

As I go about my daily work on the farm, I can't help notice that there is a natural balance in nature. That too much of one species can lead to a decline in another. Vicious corbies raiding peewits' nests is one example that I have already highlighted, but there are many others and they are often caused by man's intervention.

Man is very destructive of his environment. Rain forests have been felled, whilst cities, motorways, airports and the like now cover vast areas of prime land. Rivers and oceans have all been polluted in varying degrees. Even the air we breathe is contaminated with exhaust fumes and industrial emissions. There's no doubt things go badly wrong when the balance of nature is upset. Apart from pollution caused by industrialisation, there are other less obvious ways to upset the balance of nature.

Not for the first time French scientists got themselves in a muddle interfering with nature. In the mid-nineties French farmers protested at the reintroduction of wolves in a nature reserve in the south of France. Those who dreamed up the scheme hadn't reckoned on wolves acquiring a taste for lamb. That same daft idea is regularly suggested from time to time for the Scottish Highlands.

There can be little doubt about the impact wolves would have. They are very shy creatures and pose little threat to humans. Most of us would be lucky to catch even a fleeting glimpse of one.

Once common in Scotland, they would easily re-establish themselves and thrive in the Highlands where there is now a massive red-deer population. Packs of wolves culling older and weaker deer would improve the health and fitness of the herds, but they would also kill sheep, which would be a lot easier to catch.

Wolves were hunted to extinction before the Clearances made way for sheep, and modern Scotland is now a vastly different place.

Shepherds in those countries that still have wolves, like Poland and Romania, have to pen their sheep at night for safety. I'm filled with dread at the thought of packs of hungry wolves establishing themselves in our forests and killing sheep. Imagine the carnage and mayhem at lambing time. Let's hope common sense prevails, and this folly is never allowed to happen.

It's all too easy to upset the balance of nature.

Ladybirds are yet another of the farmer's friends under threat. Those pretty little spotted beetles feed on aphids or greenfly, a real scourge that spread diseases such as plant viruses when they suck the plant's juices. Each adult ladybird can guzzle 5,500 aphids during the summer months. Better still, a ladybird can produce one thousand hungry larvae in a season and each one can eat the same number of aphids as its parents.

Scotland's ladybirds have recently come under threat by the small, parasitic wasp, *dinocampus coccinella*. It's a flying, ant-like insect that lays its eggs in ladybirds that eventually become paralysed by the developing larvae. Scientists at the Scottish Crop Research Institute near Dundee reported back in 1999 that nearly three-quarters of the seven-spot ladybirds found in their area had been affected by the wasp.

That recent surge in the pest was blamed on global warming, which allowed the wasps an extended breeding season. Normally the wasps only manage two generations a year, but scientists reckoned they are now achieving three. As a single wasp can lead to the infection of ten thousand ladybirds in two generations, it's little wonder that by achieving three generations they were spreading the problem so rapidly.

So how important is the ladybird to farmers? The sprays that control aphids cost British farmers £100 million per year. If ladybirds died out, farmers would have to spray at least twice as much. You don't have to be top of the class at arithmetic to work out that means spending £200 million instead of £100 million.

Elsewhere, rabbits were introduced to Australia with devastating results when they took over that continent in the absence of their natural predators like foxes. Coypus from South America escaped from fur farms to destroy East Anglian riverbanks with their burrowing until they were finally eradicated by trapping. Another fur animal, the Scandinavian mink, has also escaped in large numbers and is now established as a ruthless killer.

We should never lose sight of the fact that it takes thousands of years for nature to evolve a system that's in harmony. Snag is, man imports and exports so much food nowadays that hundreds of pests, parasites and diseases are being transported around the world. Some can't adapt to the climate and die out, but others develop into serious plagues.

I'm not against scientific progress. After all, commerce has helped us enjoy far better standards of living. Nevertheless, in this rapidly changing world of commerce and science we must be ever on the alert and guard against ecological disasters.

I am currently trying to keep grey squirrels out of my SSSI. They are an aggressive import from North America that soon displace our smaller, native, red squirrels to the extent that they threaten to extinguish them in some parts. They are now established in Cumbria, Stirling and Peebleshire and we are anxiously trying to keep them out of our area. From time to time they turn up. For instance, I recently found one that had been run over and I sent it for DNA tests to find out where it had come from. A few years ago the gamekeeper shot one that came from Stirling, probably on top of a lorry loaded with straw.

Periodically we set special traps in the woods that only catch squirrel fur. They're funnel-shaped and baited with irresistible peanuts and sunflower seeds. As squirrels wriggle through for the bait they leave behind some fur so you can establish if there are grey squirrels about. There are also traps that harmlessly catch squirrels alive so that you can release the red ones and humanely dispatch the greys. Snag is, my wood is full of reds, and you would needlessly catch and release hundreds. If we ever find evidence of greys becoming established, the gamekeepers will set up hides overlooking baited clearings so they can be shot without harming the reds. That may seem harsh on grey squirrels, but it's vital we protect our precious, native red squirrels.

As I said at the beginning of this chapter, the two best times of the day are sunrise and sunset. After the day's work is finished, it's easier to enjoy sunset from a leisurely vantage point and watch the countryside bed down for the evening. To be tired enough to make the act of sitting down a sensation of real relief is a pleasure that has much to be said for it. And provided that you are not over-exhausted, but just physically in need of a rest, then the mind often functions at its very best.

I enjoy summer and autumn sunsets best. A dramatic red sky as the sun slowly dips behind the hills in the west seems to show off the drama of the local scenery to best advantage. Such a sunset may well be a shepherd's delight but it is not exclusively reserved for shepherds and everyone can enjoy it.

Rooks streaming across the sky as they head back to their rookeries give way to mallard ducks flighting onto ponds or stubble fields to feed, whilst a woodcock may be briefly glimpsed flitting out of a wood to feed. Pheasants cackle in the woods as they settle down to roost for the night. Overhead, bats begin catching moths with breathtaking precision. Occasionally a lamb will bleat for its mother and as dusk gives way to darkness there are the occasional screeches of an owl. As the last light fades and the silence of the night descends, I always feel relaxed and contented. That's why such colourful and dramatic sunsets are a shepherd's delight.

8

Down corn, up horn

Most arable farmers start ploughing the stubble fields as soon as their harvest is gathered in. Traditionally, Scotland's cereals were sown in the spring but, over the years, newer, heavier yielding autumn-sown varieties have proved ever more popular. So autumn is the time of year for the ploughman to demonstrate his craft and, make no mistake, ploughing is a highly skilled task that takes years to perfect.

The plough has to be set so that the furrows are all the same size and fold nicely on to their backs, covering everything that was once on the surface. Most important of all, the furrows have to run as straight as an arrow shot from a bow. There is something satisfying about ploughing. Many workers never see the results of their own efforts day by day, but a ploughman has the satisfaction of leaving behind his work for all to see and admire.

Like many upland farmers I gave up growing grain years ago. My land really isn't fertile enough to grow heavy crops. Worse than that, my fields are about eight hundred feet above sea level so crops are slow to grow and ripen much later. I had too many disappointments with late harvests in wet weather and decided to leave grain growing to the specialists in arable areas.

Arable farming is just as technical as any other type of farming and requires considerable attention to detail. Crops need to be regularly inspected to see if there are any problems developing with weeds, pests or diseases and to decide when to spray them with the appropriate chemicals.

Having successfully nurtured his crops the arable farmer then has to harvest them, and that can be a nerve-racking, traumatic experience. 2004 was one of the worst harvests in twenty years thanks to an unbelievably wet August. The biggest casualty was the oilseed rape. It's the crop with bright, yellow flowers.

Eventually those flowers give way to pods containing small, round, black seeds that are crushed to extract cooking oil. A combination of persistent rain and humidity encouraged the seeds to germinate in their pods, rendering them useless for crushing. A fair amount of the crop from Yorkshire to the north of Scotland was ruined and instead of being harvested was ploughed into the soil.

Wheat also suffered a similar fate in 2004 as the grain turned black and sprouted. Large areas of cereals like wheat and barley were flattened by torrential rain that had seen more than three times the average rainfall for August. What a contrast between that harvest and the previous year's, which had taken place in drought conditions and was safely gathered in by the end of August.

News reports of flash floods and roads blocked by mudslides were just the tip of the iceberg. Harvest fields were sodden or under water and that caused a lot of problems. Modern farm machinery is massive compared with twenty or thirty years ago. Those big machines allow us to deal with a field in a matter of hours rather than days. We can now make the most of every dry spell and I don't know how we would have coped that autumn without our modern equipment.

Unfortunately, heavy machines can make an awful mess of a wet field. Deep ruts are obvious and unsightly but, underground and out of sight, broken or damaged drains are a more serious problem. Many farmers had to work around wetter patches to avoid such problems. It was annoying having to look at those patches that weren't harvested, but that was far better than water running down the fields from burst drains.

Most of us work with twin wheels on the tractors in very wet conditions. Having four wheels at the back of the tractor helps to spread the weight so that equipment doesn't get bogged down easily, but when modern machines get stuck they really do get stuck. There were many dramatic scenes that autumn, in which three or four big tractors were needed to pull another one out of the mire. It was the same with combine harvesters, which were a much bigger problem to deal with.

When a tractor gets stuck you can easily attach a heavy chain to it. Most have special attachment points. For those that don't, you simply attach the chain to the front axle. Combines are a different kettle of fish because there are a lot of delicate parts at the front and few strong points on which to attach a chain. Often as not that involves crawling about underneath the combine in mud and cold water.

Another problem with the land being so wet was that tractors and trailers messed up the roads. Mud stuck to the wheels in the fields and was soon left behind on the tarmac. That could have led to accidents so it had to be regularly swept up, and this unwanted task was made more difficult after the mud had been run over by hundreds of cars. Fortunately, we now have tractor-driven road sweepers rather than brushes and shovels, another modern solution to another modern problem.

That 2004 harvest was an arable farmer's worst nightmare. Those flattened crops took longer to dry out and were difficult to harvest. Moist grain has to be dried before it can be safely stored but the soaring price of oil made it extremely expensive that year. Worse, the quality of the grain was poor and didn't fetch top prices. Malting barley, which is made into whisky and beer, deteriorated and was only fit for feeding to livestock. It was the same with wheat destined for bread-making and much of that also had to be fed to farm animals.

Mind you, one farmer's loss is often another's gain. Livestock farmers like me enjoyed lower feed prices that winter as merchants snapped up cheap grain to mix into animal rations.

'Down corn, up horn' is an old saying that reminds us cattle prices often rise when grain prices are low. Many arable farmers end up buying cattle in years like that to use up their low-value grain and convert it into valuable beef. All those extra buyers looking for young beef cattle at the autumn sales pushed up prices. It's sad to recall that I profited from the misfortune of arable farmers that year.

Having said that, arable farmers profited at my expense the following year because that sodden harvest meant there was very little straw left in store. Straw is mostly the stalks of the crop left behind by the combine in long, golden rows ready for baling. Eventually, it's either used for bedding livestock or for feeding to cattle. Straw is an important part of the winter rations of many beef cattle and hill farmers like me buy large quantities of it from our arable colleagues.

The crop of straw from the 2004 harvest was lighter than usual due to the atrocious weather and became very scarce as the subsequent long winter dragged on into a late spring. As a result it nearly doubled in price and was fetching about £90 a ton in May. That literally was the final straw!

To put that in perspective, some grain was being sold for as little as £65 a ton. In other words, a by-product became more valuable than grain for a while. Traditionally, oat straw was used as fodder and barley straw for bedding. The decline in the acreage of oats being grown forced us to use barley straw as fodder.

At one time wheat straw was considered useless, as farmers believed it had little nutritional value and wasn't absorbent enough for bedding. As a result it was used for thatching stacks or protecting potato and turnip clamps from frost. That year I was forced to buy wheat straw for bedding and even fed some to my cows. My straw merchant told me modern varieties of wheat produced nutritious straw that also makes good bedding. If my grandfather had heard that he would have turned in his grave!

When straw is expensive, resourceful farmers look for cheaper alternatives. Over the years, generations of farmers have come up with alternatives to straw for bedding. Dried bracken, chaff, sawdust and wood shavings have all been used. Valuable racehorses often lie on cannabis, not the illegal drug, but cannabis sativa or hemp. It's particularly absorbent and is also dust free. Latest new substitute is shredded newspapers delivered by the lorry-load.

At one time, straw was considered a nuisance and was regularly burnt by arable farmers. Smoke from fields of burning straw was a hazard for motorists, while the ash blowing in the wind ruined washings as they hung out to dry. So laws were passed that prevent the burning of straw.

Snag with straw is that valuable dry days are needed to get it baled before the fields can be cleared, allowing the next crop to be sown. To avoid wasting those precious dry days, some farmers fit choppers on the back of their combines. That way the straw is chopped up and spread over the field. It's then ploughed into the soil as part of the preparations for sowing the next crop. Ploughing-in straw speeds up the process of sowing, but I bet many farmers regretted it that year when the price of straw rocketed. Fortunately, no two years are ever the same in farming. The combination of a good harvest the following year and more straw being baled rather than chopped and ploughed-in led to adequate supplies and lower prices.

Most harvests follow a routine. First to be combined is the winter barley that was sown in the autumn then it's the turn of oilseed rape, followed by spring barley and finally wheat. Most British wheat is used for animal feed or biscuit making. Few British varieties are suitable for bread-making, so we import most of that type.

Scotland's main cereal crop is barley that's used for animal feed, brewing and distilling. Spring barley doesn't yield as much as autumn-sown barley, but it's still a favourite with livestock farmers who crush the grain and feed it to their cattle and sheep. The type that goes to make our national drink is called malting barley and has lower nitrogen levels than other barleys. Samples with nitrogen levels of 1.5 per cent can make substantial premiums over feed barley. Each lorry-load is carefully sampled on delivery and payment is made accordingly.

Malting-barley contracts are based on premiums above feed-barley prices, although most years a lot of squabbling takes place with merchants about prices. Farmers often feel cheated in years when feed prices are low yet the price of beer and whisky stay the same. Malt is produced by moistening barley to encourage germination and then stopping the growth by applying heat. That causes the starch in the grain to be converted to sugar that's eventually fermented into alcohol.

China is now a growing outlet for malt. It seems that many of the 1.2 billion Chinese have acquired a growing taste for beer and they can't get enough of it. Despite that, prices for malting barley remain depressed largely as a result of competition from new maltsters being built in Eastern European countries like Poland. That has led to a number of closures in Scotland and a political row over how 'Scottish' Scotch really is. Buyers of grain are also setting double standards.

Most Scottish farmers are Farm Assured. It's a scheme that guarantees consumers our produce is of the highest standard. I have been a member of Farm Assured Scotch Livestock (FASL) for some time now. Every year an independent inspector visits to check all my records are up to date. In particular, attention is paid to my veterinary records to make sure all medicines are being properly used.

After all the office work, he checks the cattle sheds are properly maintained so the beasts can be comfortable, properly fed and well cared for. Handling pens and feed stores are all thoroughly vetted.

Then he inspects all my animals to check they are healthy. Even the farm dogs are given the once over.

If everything is in order my membership is renewed and I can sell my animals as Farm Assured. That should earn me a small premium. Unfortunately there are so many different logos and brands on supermarket shelves that shoppers have become confused. How many shoppers know that the Farm Assured label means those animals have been reared to the highest standard in welfare-friendly systems?

That's why Scottish arable farmers feel let down when maltsters import barley or malt from Eastern Europe to brew cheap beer. Maltsters won't sign a contract with Scottish growers unless their grain is Farm Assured. Those contracts turn out to be worthless and the merchants often break them in favour of cheap imports. They always find overlooked clauses in the contract that allow them to reduce the price or even to reject lorry-loads of grain. Foreign companies now own more than half of the Scotch whisky industry. Biggest player is Diageo, a multinational company with its headquarters in Edinburgh. Then there are the French companies that own Chivas and Glenlivet, while Bowmore is now in the hands of a Japanese concern.

Many are now worried about the Scottishness of our Scotch. Is it right to sell 'Scotch' whisky that's been made with foreign ingredients? The growing trend to distil Scotch whisky from English malt made from English barley is beginning to alarm marketing experts. Large areas of England's arable land is fertilised with sewage sludge from the cities.

Fortunately, most Scottish farmers shrewdly resisted offers from sewage works to have sludge spread free of charge on their land. They reckoned the risks of disease and contamination were too high, but the practice is common in England. The idea of drinking Scotch made from barley that was grown with English sewage could leave drinkers with a bitter taste!

Organic is a buzzword for the modern food shopper. Instead of being produced conventionally with the aid of artificial fertilisers and weedkillers, organic food is grown traditionally using composted farmyard manure, crop rotations and mechanical weeding. Large quantities of farmyard manure or muck can induce crops as heavy as those grown with the aid of artificial fertilisers. Indeed, using naturally produced muck is common in poorer countries where artificial fertilisers are almost unaffordable. Even human sewage, or night soil as it is politely referred to, is a common fertiliser in developing countries.

There's now a strong argument for protecting the good name of Scotch whisky with PGI status. Protected Geographical Indication is an EU mechanism used to protect the authenticity of local, speciality produce. Arbroath smokies are now protected by a PGI so that inferior imitations can't be sold under that name. It's the same with Italian Parma ham, while the French have widely used PGIs to protect their famous cheeses and wines. Best example is Scotch beef, which recently won PGI status. Only beef animals that have been born, reared and slaughtered in Scotland can be sold under the label of Scotch beef. Many now believe that Scotch whisky should be made in Scotland from Scottish ingredients, like malt that was made in Scotland from barley that was grown in Scotland. Scotland will have to use PGI status to protect her favourite tipple.

It's amazing the impact the weather can have on a farming business. Cold, wet weather at lambing time can kill hundreds of lambs with hypothermia. A wet summer can result in poor quality silage and hay that leads to extra feed costs during the winter.

Droughts can lead to crop failures while a wet harvest wastes the grain and results in high drying costs and poor quality. Often as not the weather determines if there will be too much, or too little, produced. In fact, it can have a bigger impact on farm profits than political decisions or hard bargaining by supermarkets.

No wonder the weather is a constant talking point among farmers. Apart from the impact it has on profits, it also affects our moods. There's nothing better than rising early on a fine morning. Everything seems to go more smoothly when the sun is shining.

Compare that with the misery of sitting in the kitchen or office watching rain splash the puddles when you should be outside making hay, shearing sheep or harvesting grain. A miserable, wet summer fairly gets you down.

9

Tattie howking

Scotland's weather doesn't always please sun-seeking holidaymakers, but while our climate may be cold and wet it does suit certain farm crops. Oilseed rape grows particularly well in Scotland thanks to our extra daylight in mid-summer compared with further south. It's the same with swedes, or neeps as we call them, and the best crops are often found in the north-east, nearer to the land of the midnight sun.

Oilseed rape and swedes also do well because our colder temperatures help to keep them relatively free of aphids. They're a real pest down south and, as any gardener knows, spread diseases like plant viruses.

I have a friend who grows organic swedes for supermarkets on his high-lying hill farm. As they're organic, he's not allowed to spray them with pesticides, but he reckons where he grows them it's far too cold for aphids, so he doesn't have to. He also grows organic salad potatoes for the same reason and that's where Scotland comes into its own. Scottish seed potatoes are known around the world for their high health status. Thanks to our cooler climate, they are free of the viruses that can destroy an entire crop. Scottish farmers earn a premium by exporting seed potatoes around the world.

Growing crops for seed is a highly specialised job as seed grain or potatoes have to be grown to the highest standards to obtain perfect germination. They have to be free of diseases and weed seeds, and the purity of the variety also has to be guaranteed. After all, if you plant Duke of York potatoes you would be upset to see Epicure or Kerr's Pink growing as well!

Producing seed to such rigorous standards is very expensive, but can be very profitable. Round about July, officials inspect seed crops to certify them as fit for seed, and they are highly trained. For instance, they are able to recognise at least forty major potato varieties of the sixty or so commercially grown. Cereal inspectors can identify twenty-four of barley, twenty-nine of wheat and sixteen of oats.

When certifying a crop of potatoes, the inspector picks a sample area at random. If the crop is less than four hectares he samples 0.1 hectares; if it is more than that, he samples 0.2 hectares. More than three rogue plants per sample plot will cause the crop to be rejected. Before the inspection, farmers rogue the fields. Squads of trained staff walk backwards and forwards through the crop pulling out those rogue plants.

Diseased plants suffering from virus diseases, such as leaf roll, or mild or severe mosaic, are also hand-pulled. All rogues are put in a haversack and dumped at the edge of the field to be destroyed. It's backbreaking work, trudging along on a sweltering hot day surrounded by swarms of flies.

It's also amazing how rogue plants can hide themselves. A gust of wind will cause the head of a wild oat or alien barley plant to duck down among the main crop of wheat. Only when you are fifty yards up the field will it spring back up to laugh at you.

Once the roguing is completed there is always the nerve-racking wait for the final inspection by the officials. Their trained eyes miss nothing. Apart from looking for plants that are taller or smaller than the main crop, they also watch for leaves that are waxier or a different colour. That's when all the missed rogues stand up to be counted. It's worth it at the end of the day. After all, when a farmer sows seed grain he doesn't want to be sowing weeds as well.

The potato harvest used to be gathered by squads of 'tattie howkers' when I was a lad. Anyone over the age of forty will recall that all tatties were picked by hand. Squads of casual workers, women and schoolchildren were picked up by tractor-drawn trailers, vans or buses at arranged meeting points and taken to the tattie fields. It took a lot of folk to gather in the harvest and that's why there were school holidays in October. Schools were a major source of tattie howkers and that October break from school was called the 'tattie-picking holidays'. Their task was to gather the tatties that were scattered on the ground behind the tattie howker – a tractor-driven spinner that scattered the soil and potatoes in the rows, or drills as we call them.

Those squads worked in stints, or stretches called 'bits', that were usually marked off with a stick or branch. Youngster had 'half-bits' and earned half the money. Gathering potatoes into baskets or buckets and then tipping them into the trailers was gruelling work that found every weak muscle in your back. Often you felt that you would never stand straight again.

Although the work wasn't that well paid, it was a chance for a family to earn some badly needed extra cash. Gradually tattie howking has died out. Changes in benefit rules make it hardly worthwhile for the unemployed to go picking potatoes. Also, I reckon the youngsters aren't as fit as they used to be and probably wouldn't last long in the fields.

My own October school-holiday week was spent howking tatties as part of a much smaller squad. In those days, most livestock farms had a field of swedes that were an important part of the winter rations of both cattle and sheep. Some of the drills were sown with cabbages that were fed to rams, or tups as we call them. Elsewhere there might be a short drill of carrots, while part of the field might have kale. Obviously, the carrots were strictly for human consumption, although we helped ourselves to a swede or cabbage for lunch, while kale was regularly used in broth. Unlike the farmhouse garden, which was often infested with root fly, the fields grew healthy carrots and cabbages. Gardens were planted with lettuce, onions and early potatoes that were followed by leeks.

There were also several drills of potatoes in that field of swedes and it was the task of the women and children on the farm to gather them, so they could be carted to a shed and stored for the winter. Free milk and potatoes was a valuable perk for most farm workers in those days. One of the favourite varieties was Kerr's Pinks and those wonderful potatoes were the staple diet of a lot of farmhouses at the time.

Nowadays potatoes are harvested mechanically. Mighty machines separate potatoes from the soil and stones and then pass them by elevators into trailers that cart them to modern refrigerated stores. Mind you, those machines need fine weather to operate properly, and mud sticking to potatoes makes them worth less. When they're wet they don't store well and are prone to all kinds of diseases.

Prince Charles, one of Britain's best-known farmers, once had his produce rejected by large supermarkets after it failed to meet their criteria. The Prince is an enthusiastic organic farmer who found it hard to grow perfect vegetables, fit for supermarket shelves. First, his organic carrots were rejected because they were too crooked; fussy supermarkets want perfectly straight carrots. Then, organic potatoes – grown at his Home Farm, near Tetbury, Gloucestershire – were rejected because they weren't shiny enough and were covered with skin marks. Fortunately, Prince Charles struck a deal with South Gloucestershire County Council to provide local schools with 100 tons of his potatoes.

It all goes to show just how fussy supermarkets are.

Snag with organic farming is that it's harder to control diseases and pests. Don't forget that the Irish potato famine started because of an epidemic of blight in the crop. They were growing their potatoes organically and didn't have chemicals to spray on the crop when it became infected. Diseased vegetables, or those that are partly eaten by insect pests put most of us off them. There's also time and waste involved in scraping off the infected parts. No wonder scientists developed the technology to grow healthy, wholesome crops that taste as good as they look.

Prince Charles's experience demonstrated the problems with organic farming: lower yields and more waste.

In a world where many go to bed hungry, it's scandalous that we throw away good food because it's blemished with disease or pests. I read somewhere that if the whole world went organic we would need to farm three times the amount of land that we do at present. Worse, there would be a real risk of biblical-type famines as crops failed due to uncontrolled weeds, diseases or pests.

Few folk can afford the extortionate price of organic food. Most of us have tight budgets and buy more economically priced food produced by traditional farming systems.

Those who really care about the environment should concentrate on locally produced food. Think of all the fossil fuel needlessly wasted transporting food all over the world.

Nowadays we're spoiled for choice when we do our weekly shopping at the supermarket. Rows of shelves are packed with all kinds of delicacies from around the globe. Food has never been cheaper in real terms and there can be little doubt that those of us in the affluent West live in a world of plenty.

Scottish food is often harder to find than most. That's a pity, as we produce much to be proud of, and, as it's literally grown on our doorsteps, it's fresh and tasty. Few realise the extent to which our environment is being damaged by the global trade in food. Brazil's government encourages farmers to cut down vast tracks of the tropical rainforest to create new pastures for its rapidly expanding beef herd. So, cheap, imported beef from South America comes with an environmental price tag.

Aircraft that fly food into this country cause the worst kind of pollution at high altitude, but it's the lorries that transport food to British supermarkets that cause most pollution. Every year they clock up more than twenty billion food miles; that's the distance food travels to reach our plates.

A typical evening meal could easily have travelled 50,000 miles. Beef or chicken from Brazil will have travelled 5,678 miles, Egyptian potatoes 2,181 miles, green beans from Tanzania 4,704 miles or, as a special treat, asparagus or baby corn from Thailand, 6,000 miles. The side salad might contain Spanish celery, 1,140 miles, and a glass of Australian wine will add a whopping 12,000.

To finish off the meal you could have a piece of Canadian cheddar, 3,400 miles, or the healthier option of Californian cherries, 5,430 miles.

It all adds up!

Food that has to travel long distances often contains special additives or preservatives that prevent it from going off and extends its shelf life. That's one of the reasons farmers' markets are proving so popular. Shoppers are becoming increasingly concerned about the wastefulness of transporting food over long distances and now prefer to buy locally produced, fresh food.

Another way to help the environment is carefully to study the food labels to see how far the food has travelled and put it back on the shelf if it isn't British. Little wonder it sometimes takes so long to prepare a meal. In some cases it may literally have travelled halfway round the world!

One of the big problems with potatoes is that they are prone to wild fluctuations in price. The tiniest change in production of most farm commodities can dramatically affect the price a farmer gets. Produce just 1 per cent less than market requirements and you have a scarcity that leads to rising prices. Produce just 1 per cent more than is needed and you saturate the market and prices fall. Prices can vary by 50 per cent or more just because of small swings in production.

Once farmers realise that there is a shortage they often hold out for even higher prices. That creates an even greater shortage leading to ever-higher prices.

At one time farmers produced whatever they liked and then thought about how to sell it. After the second world war the government introduced a system of guaranteed prices for most of our produce. Latterly the EU bought unwanted surpluses and stored them. Until 1995 the Potato Marketing Board (PMB) controlled the area planted through quotas. Each farmer had a limit to the number of acres he could grow. If he wanted to expand, he had to buy more quotas. That system worked reasonably well, and was backed up in those years of bumper crops by a buying scheme operated by the PMB. Surplus potatoes were bought at set prices and dyed an unattractive purple before being sold as animal feed. That helped keep prices at reasonable levels.

Fortunately cattle love crunchy, raw potatoes and when a cheap bonanza is available we feed them to our beasts. They can also be mixed with grass and stored in the silage pit as winter feed.

With the PMB and other support structures long gone, increasingly we have to secure our markets before we think about production. Futures contracts are the buzzwords nowadays and apply to everything from milk and pigs to grain and vegetables. Snag with these contracts is that in a year of scarcity, when prices soar, the farmer loses out and has to sell at the price agreed earlier. However, in a year of surpluses you are guaranteed an outlet at a fair price.

Potatoes are also used as a source of industrial chemicals such as starch and the prices for those products don't vary that much, but prices do rise and fall at the greengrocer, supermarket or chip shop. Housewives are a canny lot and soon change to something cheaper. Rice or pasta is just as tasty and doesn't need peeling. Experience has taught farmers to fear that once a housewife has changed from potatoes she may not return as a regular customer.

I well remember when the price of potatoes went through the roof in the drought years of 1975 and 1976 and it took years to win housewives back.

10

We all like a drink

In today's world, water and sewerage are taken for granted. We expect clean drinking water when we turn on the tap and toilets to flush when the handle is pulled. Most folk flush the toilet or drain the bath without a second thought. It goes down the drain, into the sewer and ends up at a sewage-treatment works, but thousands of farmers and other country dwellers have to provide these facilities for themselves.

Sewage is dealt with by a septic tank. It's a simple system in which naturally occurring bugs break down sewage so that clean water overflows and seeps away. Snags can occur when over-zealous housewives use powerful detergents and cleaning agents like bleach. That kills those beneficial bugs so the system stops working. End result can be raw sewage seeping into watercourses to pollute the environment. Septic tanks are a good system, but they aren't magic boxes. As more and more townies settle in the countryside such system failures are becoming more frequent. Few even know where their septic tank is, let alone that it has to be nurtured and regularly checked.

Private water supplies are a bigger problem. Some come from springs and supply beautiful, crystal-clear water, but most are simply based on land-drainage systems or even open burns. Few folk bother to check their water supply or maintain it properly and the results of that carelessness can be catastrophic. Water-borne diseases – like salmonella, cryptosporidium and E. Coli 0157 –

often cause serious illness and can be fatal, particularly for visitors to the countryside.

Cattle and sheep droppings contaminate burns and enter the system. Unfenced water-holding-tanks, or tanks with badly fitted or broken tops, allow contamination to take hold. Moles, rats and other small mammals seeking a drink sometimes fall in and drown. Many tanks have hundreds of dead worms, small mammals and livestock droppings settled at the bottom.

The best way to be sure of a good, wholesome supply of water is to check it regularly. The whole system should be regularly cleaned out, disinfected and flushed. Before private water reaches the taps it should be filtered and passed through an ultra-violet-light treatment system to kill off bugs. To prove that private water supplies are safe you need to regularly analyse them. Most country dwellers have built up immunity to many of the bugs in their water supplies, but as I said, disaster frequently strikes visitors. That anyone should become ill is easily prevented by cleanliness and common sense.

One of the pleasures of living on my farm is the water; it is cool, crystal-clear and refreshing. Better still, it makes a grand cup of tea. I reckon there's no better cuppa in the land. Mind you, that same water supply has also been the bane of my life. Until twenty years ago we pumped it from a spring that lies about quarter-of-a-mile below the farmhouse. Those old Lister pumps had a mind of their own and invariably broke down at Christmas, New Year or at weekends; times when you couldn't get spare parts.

Now and again, in an exceptionally dry summer, the spring was reduced to a trickle, unable to supply the farmhouse, never mind the livestock. In times like that we grazed the stock on the hill where there are burns with plenty of drinking water. One very dry summer we had to cart water from a neighbouring farm for six weeks. That involved a couple of trips a day with a 250-gallon tank mounted on a trailer. It's amazing how much water the farmhouse and a few animals in the wee field nearby can get through in a day. The children loved it when there wasn't enough water for a regular bath. And when that drought finally broke it took at least three weeks of steady rain before the spring started to run again. There is nothing more soul-destroying than carting water in the pouring rain.

The previous tenant of my farm had a small dairy herd that needed a good supply of drinking water as well as enough to wash up the milking utensils twice a day and hose down the byre.

Originally, he used a windmill to pump the water but it was unreliable and he replaced it with pumps driven by a petrol engine. Eventually they were replaced with electric pumps when he accidentally spilled petrol that burned the pump-house down. Although the automatic electric pumps were a big improvement on anything that had gone before, you still can't beat gravity as a method of moving water.

After many complaints to the estate factor, my landlord, or laird as we call him, finally relented and installed a gravity system – just as the Good Lord intended water to flow. Now our water is piped a couple of miles from a burn high in the hills above us, and it's almost as pure as our spring water used to be. We connected to a large pipe, belonging to Scottish Water, which takes water from the burn to a small reservoir.

What an improvement! No more pumps to maintain.

We laid the main pipe through the centre of the farm so each field now has a water trough. Despite installing the system ourselves, and the water not being treated, it was metered and we had to pay for it. That cost about £1,000 a year, but it was worth it.

I remember an incident when a leak developed in the pipe and the quarterly water bill rocketed. Using our own water meters we eventually found the leak just below ground level at a water trough. The water had been gushing unnoticed into a land drain. It took a lot of effort to find it, and cost me £800 extra in water charges.

It seems daft to have to pay for water when you remember that Scotland has an abundance of the stuff. It seems it never stops raining. Fortunately, Scottish Water agreed to supply our untreated water free of charge a few years ago.

It's amazing how much water 2,000 thirsty sheep and 160 cattle can drink on a hot summer's day. Dairy cows are a lot thirstier than beef cows, as they're milked twice a day. The milking parlours and holding yards have to be hosed down after every milking. Water bills on dairy farms are now so large that farmers are looking for savings. One I know saved £1,700 a year by catching all the rainwater falling on the roofs of his farm buildings and pumping it up to a header tank to supply drinking troughs. Another saved £2,000 a year by locating and sorting all his underground water leaks.

Being a tenant farmer I don't own any land as my farm belongs to my landlord and I pay him a rent. All the livestock and equipment belong to me, however. Well, almost all, as I must confess some of it belongs to the bank by way of my eternal overdraft.

I also own tenant's rights. Certain fertilisers such as lime, phosphates and potash have a long-term benefit to the land. If I were to leave the farm the new tenant would be bound to pay me compensation – or tenant's rights as we call them – for the residual fertilisers in the land. Similarly, in the event of my death, my widow would claim those rights.

Other tenant's rights may include buildings, fences, drystane dykes, drains and sheep pens. For instance, the landlord might not have been prepared to build a new cattle shed, so to compensate for building a shed at his own expense, the tenant will be paid the value of that shed when he terminates the tenancy.

Landlords don't often contribute to improvements in the farmhouse. Central heating, double glazing, fitted kitchens, showers and the like are all installed at the tenant's expense. On leaving the farm, those improvements are often taken over by the incoming tenant at valuation.

Improvements and tenant's rights can be worth a lot of money, and most new tenants try to drive a hard bargain. Outgoing and incoming tenants often resort to employing costly arbiters to act on their behalf and that can lead to ill feeling; sometimes one of the parties involved will come away from the deal thinking he has been short-changed.

Now, drains are elusive creatures. Logic may dictate it should run a certain way, but when you dig three-feet down it's nowhere to be found. Dig five yards to the right, then five yards to the left, open up several more promising sites, eventually go back to the original hole, dig six inches further to the right and three inches deeper, and there it is! That in a nutshell is what draining is all about.

The same principle applies to underground water pipes in the farmyard. After thirty years' experience I now have in my mind a pretty accurate map of where all the pipes and drains are located. A map painstakingly etched on my memory after years of digging and searching. Such a map will be the reward given to the next tenant if he pays me fair compensation!

In return he will also take over a relatively modern water system, although I can't guarantee that it's completely trouble-free.

Frozen pipes in the cattle sheds can be a problem in winter. Whenever it looks like it's going to be frosty I drain the water out of the pipes. That way I only have to thaw the stopcock and inlet pipes in the morning. That's easily done by slowly pouring hot water over them.

Sometimes I get caught out and the pipes freeze during the day, and that can lead to a lot of extra work. On those occasions I stand on an old milk churn and pour cups of hot water over the pipes in the eaves above my head. Needless to say, most of the water runs back down my sleeves or drips onto the back of my neck. Not a pleasant start to the day.

It goes without saying that I've learned not to get caught out and always keep an eye on the weather, but despite all my best efforts my cattle sometimes go without water, and not because the pipes are frozen. The filter at the water meter sometimes becomes blocked and cuts off the farm supply. The first time that happened I knew something was wrong because my wife complained there was no water in the kitchen. Fortunately, Scottish Water responded quickly and had the blockage cleared by night, but that was just the start of my troubles.

My silage was very dry that winter thanks to a glorious summer the previous year. Cattle on a diet of dry silage get very thirsty after going all day without a drink. They are also very impatient, so when the water started to flow again they fought each other for a drink and damaged the water troughs.

Cows weigh nearly half a ton; so half-a-dozen cows fighting round a trough means there is nearly three tons of muscle pushing against it. To avoid all that, I tie up the ballcocks in the troughs with string. Then I undo one ballcock at a time, to let the water flow again, and stand next to the trough to make sure they queue in an orderly fashion with the best of manners. Once the pen on either side is satisfied, I move to the next trough

After a couple of hours, every beast had drunk its fill and I could head into the house for a well-earned dram to quench my own thirst.

11

The golden fleece

Once the grass starts growing again in the spring most British sheep start partially to moult their fleeces and a gap between the old fleece and the new one develops, called the 'rise'. Shearing, or clipping, sheep can begin when the rise is about an inch, as that's enough room for the shearer to comfortably work his shears between the old and new fleece. Waiting for that rise to develop can be a frustrating time.

Some ewes run about with parts of their fleece hanging off, while others have completely lost theirs. Everywhere small bits of wool lie about, making fields look untidy. Worse than that, sheep get itchy at that time of year, particularly in warm, showery weather. That's when they love to roll over onto their itchy backs for a good rub on the ground. Unfortunately, because of their shape and the way a heavy fleece spreads out, they sometimes can't roll back over to get onto their feet again. They may look ridiculous lying on their back with their four hooves helplessly pawing at the sky, but their predicament can be fatal. You see, 'couped' sheep, as we call them, can't chew their cud properly. That leads to a deadly build-up of methane gas in their stomach, or rumen, which kills them. Fortunately, the cure is simply a matter of rolling them back over so that they can get onto their feet and patiently holding them steady until they regain their balance.

In humid, showery weather sheep have to be checked regularly for coupies, as often as three times a day, and that hassle invariably precipitates shearing. Shearing is best described as separating one woolly jumper from another! Those pesky sheep love to wriggle and squirm during the two-minute process.

Top shearers are super athletes. They need the eyes of a skilled marksman and the stamina of a marathon runner. Working bent-over double and shearing a seventy-kilo sheep is backbreaking. Experts make it look easy, but, from experience, I can tell you that sheep really struggle if they aren't held properly. It's reckoned that a day's shearing is the equivalent of running a marathon.

Good shearing technique involves holding the animal firmly but comfortably so that the fleece is removed in one piece. Easier said than done! Shearing was never my strong point and the best I could manage was about 120 a day, compared to the 400 or so a professional can manage. My miserable tally was achieved after much sweating and backache. Bending-over double all day long isn't a natural posture and eventually leads to back problems. Occasionally, ewes would struggle and catch me off balance, forcing me to switch off the electric shears so that I could restrain the struggling sheep.

The idea is to have the sheep sitting comfortably on one of its hips supported by your legs, and first remove the wool from the brisket and belly. As you work round the body you move the sheep slowly into different positions so that it never becomes uncomfortable or restless. Finally, it should run away leaving the fleece behind in one piece ready to be rolled and packed.

Inexperienced shearers, because they make the animal uncomfortable, get caught off balance when the sheep unexpectedly tries to wriggle free. The more you struggle to control them, the more tired you get. Tired shearers find it even harder to hold their sheep properly.

Holding the sheep correctly also allows you to work the shearing hand-piece properly. A badly held sheep allows its skin to wrinkle rather than being kept taut, and that can lead to cuts.

Those nicks and cuts are often no more painful than the ones I get when I shave in a hurry but they can attract flies, and that leads to open sores. In such cases we apply Stockholm tar to the wound to keep the flies at bay until it has healed.

Shearing, like making hay, depends on dry weather because wet fleeces don't store properly. Fortunately, sheep fleeces dry remarkably quickly and if you are in a hurry to get started after a shower it's simply a matter of standing a group of sheep on top of an exposed part of a field till the breeze dries them.

Many take pride in their ability to shear and the best take less than a minute to give a sheep its annual short-back-and-sides. Their prowess is tested at shearing competitions, which are regularly held during the summer months. The best of these events took place when Scotland hosted the Golden Shears world shearing championships in 2003 at Ingliston, Edinburgh at the Royal Highland Show.

The Golden Shears championship was established in 1980 but 2003 was the first time it had been held in Scotland. More than one hundred of the world's top shearers from about twenty different countries competed for the title. The event was held in the specially constructed MacRobert pavilion at the Royal Highland centre. It's a purpose-built facility that cost £300,000 and was reckoned to be the best in the northern hemisphere. When it's not being used for shearing it doubles as a venue for rock concerts. An Olympic Games-style village was established for competitors at nearby Oatridge Agricultural College, Broxburn. Mind you, the organisers of the Olympics don't have to supply their competitors with four thousand sheep, which came from farms in the Borders and were transported to Ingliston every day. Watching those world-class shearers was one of the most exciting events I have ever experienced.

In common with many farmers, I am now on the wrong side of fifty and no longer fit enough to shear properly. Indeed the average age of Scottish farmers is now fifty-nine and rising, whilst most of the one thousand or so full-time shepherds left in Scotland are a similar age to me. That's one of the big problems in sheep farming around the world. The workforce is ageing as young men seek easier, better-paid jobs elsewhere.

The average Australian shearer is now in his fifties and there are few young men willing to replace them. It's the same in New Zealand where the average shearer is now in his forties. So the race is on to design new shearing techniques that aren't as physically demanding and don't require the shearer to work all day with his back bent-over double.

About ten years ago a New Zealander called Greg Moffat invented the Moffat shearing table that allows shearing whilst comfortably standing upright and using standard shearing equipment. Sheep are laid on their backs on top of two rollers set at waist height and then have all four legs secured by ropes in such a way that it is stretched and firmly restrained. The shearer then rotates the sheep by operating a foot pedal that rotates the rollers.

Most of the folk in New Zealand who use it find they can easily shear 150 a day whilst standing comfortably. I imported one, but found a major drawback: getting the sheep restrained on the rollers is often a bit of a wrestling match and really needs two people.

Both Australia and New Zealand are currently designing races with automatic-catching devices that will overcome that problem. There are several different prototypes being developed and it is hoped the new equipment will be perfected and commercially available in a few years' time. Such equipment can't come soon enough for the Australians, as they are anxious to extend the careers of their ageing shearers. Making shearing a pleasant job instead of an athletic challenge could also encourage more youngsters to have a go.

Once the sheep is shorn we roll up its fleece individually and pack it into large sacks, or sheets as we call them, that hold about thirty to forty fleeces depending on the type of sheep. Traditionally, they were suspended from ropes and children were often given the job of trampling the wool down into the sheet. The wool is then sent to one of the depots owned by the British Wool Marketing Board (BWMB) where it is graded.

Different grades have different end uses and fetch different prices at auction. The BWMB pays farmers the average price for each grade it has sold during the year. Incredibly, there are about 130 grades. That's due to the large number of different breeds of British sheep and all the permutations of cross breeds that can result. Little wonder wool-graders are highly skilled and have to undergo years of training.

Wool prices vary from year to year and range from virtually nothing to over £1 a kilo, although the average is slightly better than fifty pence. With an average fleece weighing just over two kilos that leaves a cheque of about £1 a head, and that is barely enough to cover the costs of shearing. Man-made fibres have forced prices down. Sheep farmers love to moan about the falling price of wool and how at one time it used to pay the farm rent and now doesn't even cover the cost of shearing. Another favourite moan is that a suit containing a little more than two kilos of wool can sell for over £300! As a farmer once quipped to me, 'Wool is that bad a price that I can't afford to buy a new wool suit.'

A good way to save the planet from global warming is to wear woollen clothing to keep warm. Think of all the energy that's wasted around the world by turning up the central heating. Offices and homes are so hot nowadays it's unhealthy. No wonder we get so many colds after working in sweltering offices without any windows opened for ventilation. Folk wander about their offices scantily clad in shirtsleeves, a luxury the planet can ill-afford. Imagine the amount of fossil fuel that's needlessly squandered to generate all that heat.

It's not so long since office workers wore wool suits or tweed skirts to keep warm. Back home we wore jumpers and cardigans because the cheapest way to keep warm is to wear woollies.

Main picture: The author with his faithful sheepdog, Tweed, on Auchentaggart, a typical hill farm in south-west Scotland. The picturesque Leadhills are in the background.

Inset: The farmhouse at Auchentaggart.

Wedding bells. Like all newlyweds Carmen and I were full of hope for the future.

This magnificent Clydesdale, called Billy, was very similar to the last working horse on Auchentaggart. The photograph was taken in 1938 during hay-making on a neighbouring farm.

Top: A ewe and her twins. The mother is licking her newborn lamb.

Bottom: Orphaned lambs have to be hand-reared until they can be fostered onto a suitable ewe.

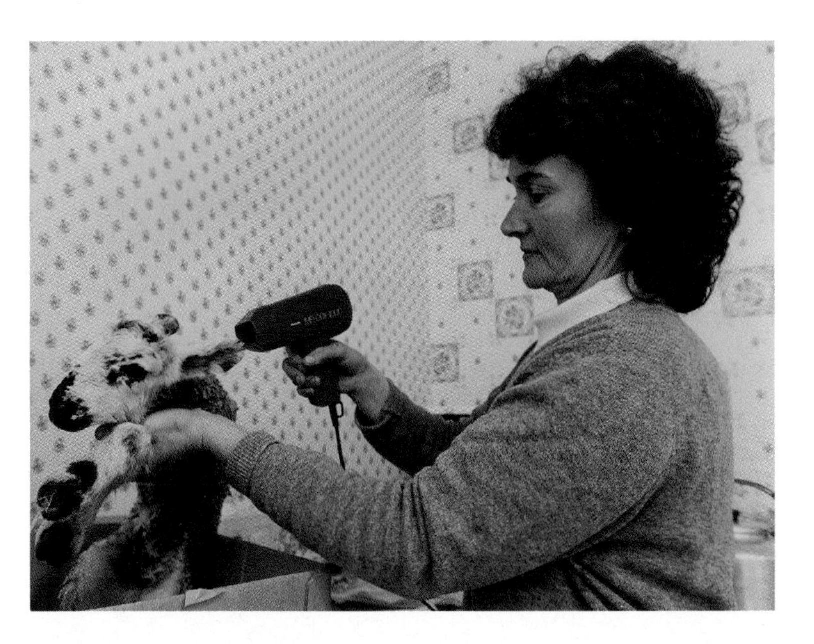

Weaker lambs need to be warmed up to recover from hypothermia. Carmen is not really giving the lamb a wash-and-blow-dry! (**top**)

What a team! The key to success in farming is your 'better half'.

The pecking order. **Top**: The delicious eggs from those free-range hens made home-baking taste so much better. **Bottom**: The male turkey is fluffing up his feathers to protect the hens. Our kids were terrified of him; he would attack anyone who came too close.

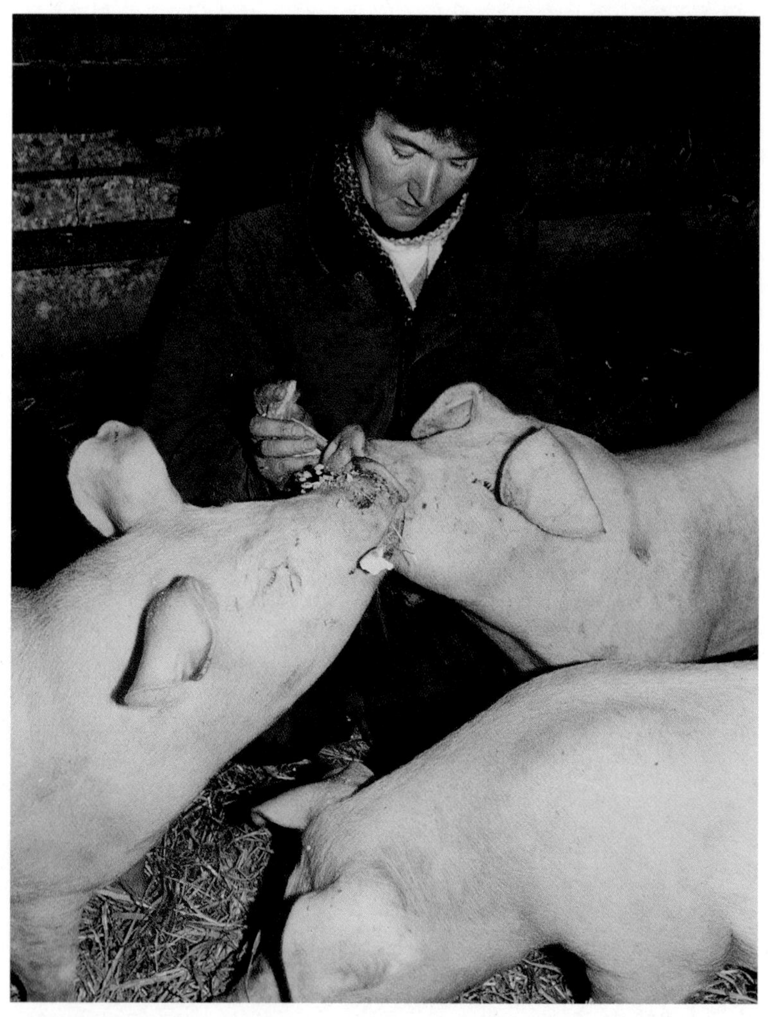

Give us our daily bread. Carmen collected stale bread from the baker every day to feed our ravenous pigs. Thanks to infernal government regulations this practice is now illegal.

As a lad, I often woke on a frosty morning to marvel at the patterns Jack Frost made on the windows. How many children get that chance nowadays in their double-glazed, centrally heated homes?

Despite the cold weather outside, we slept snug under our woollen blankets and after breakfast went outside to play, warmly wrapped up in our winter woollies.

Wool is unique because it also warms up when it gets wet, thanks to a harmless chemical reaction between water and wool fibres. That's why country folk like gamekeepers prefer to wear tweeds when working outdoors in cold, wet weather. They may get soaked to the skin, but their tweeds keep them warm and snug, unlike other waterproofs. No wonder sheep prefer their woollies during the winter!

Wool at one time made this country great. Profits from spinning and weaving enabled Britain to become a powerful industrial nation and to found an empire. Indeed the Lord Chancellor sits on the woolsack to this day in recognition of the historical importance of wool to the nation.

Fleece weights vary as widely as the price. Fleeces from little Shetland ewes can weigh less than a kilo, while those from a Devon Longwool can be seven or eight kilos. Fine wool tends to come from hot countries like Australia that is famed for the fine-wool Merino breed.

Finer wool is used to weave worsted cloth, Harris Tweed or is used in the knitwear trade. Other types of wool are used to fill mattresses like Japanese futons. Because of our wet climate British wool tends to be coarse and around 70 per cent of it is made into high-quality, hardwearing carpets. British wool is world-famous for being bulky and that gives a carpet a luxurious, springy feel.

The BWMB has three depots in Scotland: at Evanton, Irvine and Galashiels. It's fascinating to watch wool being handled, graded, baled and then prepared for auction at sales held every fortnight in Bradford. Too many of us forget that when the wool leaves our farms it's only at the start of a long manufacturing process of grading, scouring, blending, combing, spinning and weaving before it ends up as clothing, carpets or bedding.

Galashiels, for instance, handles three thousand tons of wool annually from about 1.5 million sheep in the surrounding area. It's down to just three graders to handle that massive amount of wool. Each one inspects half a million fleeces every year before putting them in the appropriate skip to be taken away and baled. I bet they don't have to count sheep to get to sleep at night!

Mind you, others employed by the BWMB used to have to count sheep. The BWMB is a massive organisation that handles the wool from about sixteen million breeding sheep. That comes to a hefty thirty-six million kilos every year and by law all of the sixty thousand sheep farmers in Britain must sell their wool to the Board. The board of directors is mostly made up of farmers, who are elected by registered wool producers, but this is no ordinary election and at one time sheep had the right to vote!

Each farmer has one vote, but at one time every ten sheep he owned counted as another vote. That way a farmer with 1,000 ewes would cast 101 votes. So instead of spending time at markets canvassing farmers, candidates in an election should maybe have gone to sheep gatherings as well. After all, ignoring the sheep vote could have led to a ba'ad result on the day of the election! Nowadays, instead of every ten sheep being worth a vote the system has been changed so that it's based on the weight of wool sold and every fifty kilos of wool is worth a vote.

Many of the poorer types of wool are virtually worthless and that's prompted the BWMB to look for new uses. One brilliant idea was to use it as lining for hanging baskets. The wool retained moisture so that gardeners didn't have to water the flowers as often. Other alternatives included insulation for lofts and using wool as a foundation layer for hill paths, to help bind the rubble laid on top.

Manufacturers have also come up with an amazing new innovation for the home, based on wool. But it's not for humans; it's strictly for the residences of our feathered friends. Building nests can be a time-consuming business. Watch how rooks go about refurbishing their nests and turn that process into a seriously competitive affair. As one flies off to find another twig, others craftily steal the twig that has just been laid rather than fetching their own. Some ground-nesting birds, like peewits and curlews, get by with a few bits of grass and straw that blend in with the background, whilst others, like busy swallows, labour long and hard to build their nests with mud. Most small birds spend a lot of time lining their nests with warm materials like feathers, dry moss, fur and wool. That helps to keep the eggs and chicks warm and dry.

It's easy enough for country birds, as such materials are literally lying everywhere, but spare a thought for their city-dwelling cousins. Wee birds in Edinburgh, Glasgow or Dundee have to fly a long way to find a piece of wool hanging on a fence.

To overcome that, the RSPB used to advise city dwellers to empty their vacuum cleaners in the garden. That way, city birds could find a supply of dog hair and fluff. Then, quite recently, manufacturers came up with their brilliant idea to help nesting birds. They designed a metal spiral and a cardboard box that can be filled with wool and hung up in the garden or from window sills. Birds simply help themselves to pieces of wool and both devices can be refilled. The idea has the backing of the RSPB and should make life easier for our feathered friends.

12

Where's the beef?

British beef farmers were forced radically to rethink their production systems, and became world leaders in concepts such as quality, traceability and production-assurance schemes. It's just a pity that it took an event as traumatic as the BSE crisis to make us take a long, hard, critical look at our industry and to realise that it had all gone horribly wrong.

At one time, British cattle breeders were the world leaders. Breeds such as the Aberdeen Angus, Shorthorn and Hereford were exported all over the world to found beef industries in North and South America, Australia and New Zealand.

The reason that those breeds were so successful and suited those new-found plains was that they had been developed to convert grass into cheap, tender beef, and I don't use those terms lightly. Such beef has fine, textured meat with a marbling of fat, allowing it to retain moisture and flavour during cooking.

Sadly we lost most of our breeding advantages through an unfortunate set of circumstances when we joined the European Union (EU).

Thanks to our hill and upland regions in Scotland, Wales and England, we had developed a specialist, suckler-beef industry, where calves are reared naturally suckling their mothers and then fattened extensively on grass. Traditional, hardy hill-breeds such as the Galloway, Highland, Luing or Welsh Black were crossed with 'improver' breeds such as the Shorthorn to produce beef cows that were ultimately crossed with 'terminal' beef sires such as the Aberdeen Angus or Hereford.

Developed over the centuries, this brilliant system combined the hardiness and mothering abilities of hill breeds with the beefy characteristics of the terminal sires. The snag was that most traditional breeds involved were slow-growing and produced lightweight carcasses that became over-fat when pushed to heavier weights.

Continental Europe's beef industry developed along different lines. There the focus was on intensively fed dairy bulls or crossbreeds from dairy cows after mating them with the massive Charolais, Simmental or Limousin. Such cattle grow rapidly to heavy weights without getting fat and, unfortunately, EU beef subsidies were designed for that type of cattle and not our smaller, traditional British breeds. In the 1970s, British farmers rushed headlong into the new imported breeds. Herefords, Angus and Shorthorns were displaced by a whole host of European breeds.

There were teething problems. Cattle such as the French Limousin that had been bred to be fattened on intensive cereal-based rations have small stomachs and literally couldn't eat enough of our hill grasses to thrive. At the same time our small beef cows had difficulty giving birth to their monstrous calves; some were almost twice the weight of our native breeds. British cattlemen soon bred those unwanted characteristics out and refined those Continental beasts. On the plus side, we were breeding big, fast-growing, lean crossbreds that maximised our subsidies when sold into EU-intervention stores.

I can hear you all screaming how wrong that was, as we had stopped producing beef for consumers. The new name of the game was to get as big a carcass as possible into the EU's frozen stores. Years later much of that beef was given away to pensioners or sold at knock-down prices to hungry Eastern Europe. Things were to get worse. Having displaced our traditional breeds, we no longer had a supply of suitable females for breeding purposes.

Further deterioration in beef quality came about when British farmers started buying crossbreds from the dairy herd that were artificially reared on milk substitutes and high-protein, cereal-based feeds. Narrow, lean Holstein dairy cows may be superb milk producers but they do absolutely nothing for beef production. The Holstein influence on beef production was dramatic. Too many commercial beef cows were narrow and 'un-beef like', reflecting their Holstein origins.

So, in the space of about thirty years, we had developed a beef industry based on extreme dairy crosses, whose origins and methods of rearing were often unknown, that were finally crossed with massive Continental-type bulls to produce inedible beef that went into the EU's intervention stores.

Our crowning achievement, though, was the development of a new system we called silage bull beef. Traditionally, we castrated our cattle, unlike the Europeans who leave them entire as bulls. Bullocks grow more slowly than bulls and lay down fat at lighter weights. Bulls fattened intensively on cereals, and slaughtered at fourteen months, produce lean meat ideal for some export markets and adequate for the home market as mince. On the other hand bulls reared on cheaper, grass-silage-based diets take longer to develop and grow to massive weights before they are ready for slaughter.

We had created the ultimate! A massive, dark-coloured, sinewy, tough, totally inedible carcass ideally suited for long-term storage in an EU-intervention frozen store. Thank God for BSE!

Fortunately, the abolition of production subsidies and new economic pressures are now encouraging us to slaughter cattle at lighter weights and produce beef that the public will enjoy eating and pay a premium for. In addition, lower prices mean that many intensive finishing systems based on bulls aren't as profitable, so traditional pasture fattening at eighteen months is again fashionable. It's amazing just how much beef was slaughtered at close on thirty months of age in pursuit of big carcasses and subsidy payments. No wonder our beef had become so tough.

Many farmers are now terrified of buying-in diseases from other farms. Traditional breeding policies are once again fashionable and everyone wants to breed their own female replacements. My one and only BSE case occurred sixteen years ago in an Aberdeen Angus-cross-Holstein cow that had been bought as a young calf from a Cheshire dairy farm. That incident made me realise that the system was failing and forced me completely to change my production methods.

I bought traditional Beef Shorthorn bulls and crossed them with my entire herd of Angus-cross-Holstein cows for five years. The resulting Shorthorn-cross cows were then mated with Angus bulls and produce cattle that are castrated and grass-fattened as traditional eighteen-month steer beef. My current system now perfectly mimics the one that my great-grandfather operated over a century ago. The cattle I produce are sold at a considerable premium through the Aberdeen Angus-certified beef scheme, that not only assures customers of high quality, but also that such beef has been produced to the highest standards of animal husbandry.

For years farmers weren't told the ingredients of the cattle feed they purchased as such information was deemed commercially sensitive. Now it is a legal requirement for feed manufacturers to declare all ingredients and the use of meat and bone meal – which spread BSE throughout the national herd – is banned.

Barcoding in advanced slaughterhouses should enable every steak, roast or pack of mince to be traced right back to my farm. I can now tell you exactly when and where the animal the meat comes from was born, account for its sire and dam and explain in detail its medical history, movements and production system.

Assuming it has been humanely and properly slaughtered, adequately hung, skilfully butchered and cooked with flair and creativity, such beef will be a gourmet experience for the luxury end of the market. So, it's over to you chefs!

13

Farmers markets

Farmers are weak sellers and seldom get a fair price from the big buyers who dominate the market. No wonder we get angry!

Not many folk noticed when some farmers held a national strike in November 2005. That was just as well because, like a lot of farmers, I thought it was a daft idea. How can farmers go on strike?

All animals kept indoors – like cattle, pigs and poultry – have to be fed every day. It would be cruel not to. Eggs have to be gathered and packed or the system would soon block up with eggs. Cows have to be milked twice a day. Again, not to do so would cause a lot of stress and cruelty.

Even more pertinently, how do you stop crops from growing? Unlike a lot of other folk, most farmers can't withdraw their labour for even a few hours. Of course, those silly farmers who went on strike realised that. Instead, they stopped selling produce for a few days, and that's why nobody really noticed.

Daftest of all were the dairy farmers who foolishly spread their milk on the land rather than sell it. That didn't affect consumers one bit, but I bet their bank managers were furious at them for throwing money away.

Those farmers who stopped selling their potatoes, cereals, cattle or sheep had the same impact. Other farmers simply sold a little more and were glad of the chance. I can understand the frustration of those who went on strike. It followed on from demonstrations that farmers held over low milk and beef prices.

We all felt hard done by and wanted to take action, but, sadly, strikes and demonstrations are more or less a waste of time. The problem with farmers is that we are price-takers rather than price-makers. Thousands of us individually bargain with big companies like supermarkets and find that we have to accept their terms.

Others, like me, go to markets and basically stand in the auction ring with our cattle or sheep asking, 'Please, sirs, what will you bid for my stock today?' Instead of demonstrating or striking, we should form ourselves into co-ops that can sell our produce collectively from a position of strength.

Of course, that's not the way British farmers work, despite the fact that our European counterparts have developed massive, successful co-ops. We are individuals who love to kid our friends and neighbours that we got a better deal than they did. While we delude ourselves that we drive a hard bargain, our buyers are laughing at us behind our backs. From an average basket of food costing £37, the farmer receives just £11.

Milk is a prime example. Dairy farmers are paid between sixteen and nineteen pence per litre, depending on the time of year, yet it costs fifty-four pence per litre in the shops. Arable farmers get around £95 per ton for potatoes that supermarkets sell for £550 per ton, whilst carrots fetch around £80 per ton ex-farm and are then sold for £700 per ton in the shops.

It's the same with prime lambs that fetch about £50 each, or £2.50 per kilo. Butchers sell cheap flank that's made into sausage meat for about £2.60 per kilo while tasty chops sell for a hefty £9 per kilo. However, I acknowledge that it is not as lucrative as it might seem for the shopkeeper: not all of a lamb carcass can be sold; there's waste in the form of bones and fat; and butchers have overheads like rent, rates and wages.

Scotland has almost a quarter of the United Kingdom breeding flock of sixteen million ewes, yet Scots ate less than 5 per cent of the 360,000 tonnes of lamb that is eaten in Britain in a year. The average Scot ate less than four kilograms.

The Scots have never been keen on eating lamb, despite having some of the best in the world. Think of Scotland and you conjure up images of Scottish Blackface sheep, those hardy hill ewes with horns. Scottish lamb, naturally reared on heather-clad hills, is considered a delicacy and enjoyed by millions throughout Europe, yet most Scots turn up their noses at it. We prefer beef, or imported pork or chicken for a change. Sounds daft to me, but that's the way it is. For our own part, my family eats half a dozen of my own lambs every year.

Apart from the chops that are easy to grill, most cuts of lamb don't suit our modern, microwave, fast-food lifestyles. It takes time to roast a leg of lamb and not everyone can sharpen a carving knife properly or knows how to carve a shoulder joint. That's a pity, because you simply can't whack roast lamb. Cheaper cuts of lamb make wonderful stews and casseroles; like hotpot, one of my favourites.

For various reasons those who still buy lamb tend to be older and the habit is dying out. Most Scots only ever eat lamb as a curry, kebab or as a special treat in a restaurant. Scottish lamb is exported to countries like England, France, Germany, Belgium and Switzerland to name but a few. Fashions come and go. Scots holidaymakers try all kinds of exotic food when they're abroad and then continue to eat them on their return home. Many now regularly tuck into rice, pasta and couscous instead of potatoes. Look at all the different restaurants that range from Chinese and Thai to Indian and even Mongolian. If only we could persuade younger folk to try lamb, I'm sure they would like it and want to eat it on special occasions.

Beef is much easier to sell in Scotland although farmers don't always reap the benefits. It takes me about three years to produce a beast from conception to slaughter, yet a butcher can sell it within a couple of days. I get about £1.90 per kilo while supermarkets sell beef for an average of £4.25 per kilo. Prices in the shops range from £4.80 for steak mince to a whacking £11 per kilo for fillet steak.

Butchers reckon on a 60 per cent margin on the beasts they sell. So a prime beast that took me three years to produce might fetch £550 at market but will probably be sold by the butcher within a few days for more than £900. As a friend once observed, 'they have their cash in the till that night but I have to wait three weeks to be paid'. No wonder many farmers have started selling direct to consumers through farmers markets and farm shops. That way they receive the full price for their produce as well as giving shoppers the chance to buy quality fresh food direct from the farmer who grew it.

Farmers markets were a feature of the Great Depression of the 1920s and 1930s; they have reappeared in Scotland and are thriving. At one time the wives loaded up a pony and trap with produce and went to the weekly market in the town. Selling from stalls, their takings were a badly needed supplement to the meagre farm incomes of those hard times. That tradition fell by the wayside with the advent of marketing boards and guaranteed prices, but, over the years, farm produce has gradually fallen into the hands of powerful middlemen and supermarket buyers who all take an ever-increasing cut. Not surprisingly, farmers have become disillusioned with a system that pays only a fraction of the end price and have reverted to the old ways of selling direct to the public.

Bristol farmers market, launched in 1988, is now one of the most successful in Britain. Others followed that lead until finally the first Scottish one was held in Perth city centre in April 1999.

There were fourteen stalls at that first market selling meat, vegetables, frozen soft fruits, wines, plants, home baking, fish, eggs and preserves. Some of the goods were organic and all had originated within the Perth and Kinross area.

In addition to locally produced beef, lamb and pork, ostrich and wild boar were on offer to tempt the more adventurous. Some had completely sold out by lunchtime. One ecstatic farmer sold the equivalent of a side of beef, a pig, two lambs and forty pounds of bacon. 'I could have sold the same again if I'd brought it,' he enthused. Another exclaimed, 'I'd rather move an electric fence with a herd of cows behind it than try to control the customers who mobbed the stall. At one stage there were hands coming from everywhere.' One of the most pleasing aspects for those involved was getting the full retail price on the day, instead of having to take the paltry sums offered by middlemen.

Those with established farm shops and mail-order businesses found a host of new customers. Snag was, a long list of strict regulations on food hygiene had to be complied with. There was also the new experience of dealing directly with shoppers, which gave the farmers and their wives the chance to explain how the food had been produced.

Sadly, people are losing touch with the land. Not so long ago, I'll wager many of you had farming connections. You might not necessarily be related to a farmer but most of you knew one or two. Some of you would have worked on farms for extra money during school or college holidays or to tide yourself over a spell of unemployment. Others got involved through the wartime efforts of the Land Army. Whatever the connection, there are few people past their middle age without some knowledge of farming life.

Today's youngsters are different. Born in a modern city, they've little reason to venture into the countryside and parents are unhappy about youngsters cycling on today's congested roads.

Town-centre leisure facilities, televisions and video games mean few of our young people are tempted to head into the country. That leaves us with a generation that knows little or nothing either of country life or how the food they eat is produced. Some modern children find it hard to believe that milk comes from a cow rather than a carton in the fridge. Do they make the connection between piglets and bacon? Do they realise that their morning rolls and biscuits come from wheat?

I used to watch my children bring their town friends home to play. I have watched the wonder on their faces as they collected warm eggs from under a hen. Most children love animals. Once they become accustomed to the smells of a cattle shed or pigsty, they love to touch livestock. There is nothing quite as enjoyable as watching a little girl cuddle a young lamb for the first time. Stroking a pig's back, letting a calf suckle your fingers or holding a young chick are all exciting experiences for a city kid.

At one time townsfolk themselves kept livestock. Not so long ago it wasn't unusual to have hens in the backyard. A pigsty at the back wasn't uncommon for a town house. They even kept cows. After the early-morning milking those cows were herded up High Street by the town drover to the common grazings. Later in the evening the drover would bring them back to High Street to be collected for milking in their stall at the back of the house. All this meant that even town children had a lot of farming knowledge.

Sadly, that has all changed and it's a change for the worse, because today's children are the farmer's customers in the future.

I believe that it's important we all know how our food originates. There would be fewer food scares if more folk realised the care that went into producing it.

This country is no stranger to food scares. Salmonella in eggs, listeria in cheese, E. Coli and, of course, BSE have meant massive losses for farmers and food processors. But every cloud has a silver lining: the result of remedying these problems is that Britain now leads the world in setting new, high standards for farmers, slaughterers and processors.

Edwina Currie's famous outburst on salmonella in eggs triggered what was arguably the first major food scare. Sales plummeted to be replaced by imports containing even higher levels of salmonella. In a desperate attempt to salvage the British egg industry, the government introduced compulsory testing for salmonella backed by a ruthless slaughter policy for infected flocks. While that scheme hasn't totally eradicated salmonella, it has given us much lower levels compared with other countries.

As in all other EU countries, the use of growth-promoting hormones in livestock is illegal, yet it's commonly practised by beef producers elsewhere.

Britain also leads the way in setting high standards of animal welfare on the farm, during transport and at markets and slaughterhouses. Tethering sows is now illegal in this country, and this has led to a cost disadvantage compared with competitors like Denmark that still allows this cruel practice. The British producers who now let their pigs run free-range outdoors, or in straw-bedded yards, feel aggrieved at supermarkets selling cheap, intensively produced pork.

It's the same with poultry. Here in Britain there has been a massive move from battery cages to free range or barn systems. That's great for British hens, but it has led to higher labour and feed costs for the farmer.

Arable production in Britain, as I have said, is also of a higher standard. In common with our livestock farmers, they have established assurance schemes where independent assessors inspect farms to ensure good practice and hygienic production systems.

In common with the rest of the EU, British farmers have not been allowed to grow genetically modified crops to any great extent, putting them at a disadvantage with American and Australian competitors. In addition, British supermarkets, the dominant buyers of home-produced food, have set high standards for British suppliers; much higher than in comparable countries. Next time you go shopping, look at the label and buy only British food. It's simply the best!

A lot of farmers are coming round to the view that we have to get closer to our customers. We want people to see a herd of cows being milked in modern, hygienic, milking parlours. To realise just how sophisticated the technology in a combine harvester is. We want them to watch vegetables being graded, washed and packed in a surgically clean packing-shed. To understand the technology and welfare involved in producing thousands of eggs every day, or hundreds of pigs a week, and to talk it over with the farmer.

Supermarkets have transformed the way we shop, but are you aware they've also had a profound effect on the way we run our farms? Conveniently sited, with plenty of parking space, we can easily stock up at the supermarket for the week, and what a range of products and choice on offer!

As a wee lad I often had to fetch the 'messages' from our grocer. I stood patiently in a long queue whilst the assistant gossiped with each and every customer. When it came my turn she would take my line and fill a shopping bag or cardboard box with goods from the shelves. Sometimes she'd put a pencil mark against an item on the list and declare: 'Tell your mother we've run out, but we're expecting more in next week.' That never happens in a modern supermarket, as they never seem to run out of anything.

Thousands of different products from all over the world are stacked high to tempt even the most miserly. And look at the quality! Gone are the days of complaining about mouldy carrots or bruised potatoes. Everything is perfect. That's the way it should be, but only supermarkets have the buying power to insist on the best. In their search for the holy grail of perfect quality they have literally moved on to the farms and are dictating to farmers how they want food produced.

When they say they want their potatoes no smaller than sixty-five millimetres, and no bigger than seventy millimetres, they mean it. If a consignment doesn't come up to standard, it's rejected. If they ask for pigs to be grown a certain way they will come onto farms and inspect the systems. Once again, failure to comply leads to rejection. It's all down to what the supermarket bosses term 'due care and diligence'. They rightly reckon the housewife is entitled to good and wholesome food that's also attractive, and they mean to see that's what is supplied.

Truth to tell, supermarkets may have encouraged greater choice and better quality, but I doubt if your grocery bill is that much less as a result. You get what you pay for. If only 60 per cent of a vegetable crop comes up to supermarket specifications, then the farmer needs premium prices to make up for the 40 per cent of waste. Similarly, he needs to be properly rewarded for his massive investment in building sophisticated grading lines. So next time you pop a bag of washed carrots, a neatly trimmed bundle of spring onions or a cellophane-wrapped lettuce into the trolley, think of all the care and attention to detail by the farmer who grew and packed it.

Mind you, supermarkets are notorious for driving down farm prices and one of their tricks is 'down averaging'. Supermarket buyers don't deal at auction markets. They prefer to buy carcass meat like lambs direct from slaughterhouses. They offer a premium payment for top-quality lambs. That way, farmers are encouraged to sell their best lambs to them with the poorer ones going to auction. Unfortunately, the average reported price nationally is based on returns from the auctions. Since those returns are calculated on the poorer lambs, it means the supermarkets are basing their premium on prices that are below the true average for lambs. That leads to a downward price spiral because the supermarkets can then reduce their premium prices. This in turn has a tendency to force down the price of the lambs they don't want and, the following week, premium prices go down again. While practices like that may be good for supermarket profits it's unfair to everyone else. There must be a way of persuading supermarkets to pay farmers fairer prices. Until that happens direct selling at farmers markets will continue to increase in popularity.

One local farming couple started selling their organic beef, lamb, poultry and pork through various farmers markets. That encouraged them to open up a farm shop, which has gradually built up a successful trade. They specialise in well-hung beef. Hanging beef in a chill for three weeks or so helps to tenderise it and is the secret ingredient in good meat. Well-hung beef can turn as brown as mahogany, but it's delicious. Beef that has been butchered within a few days of slaughter is pink or bright red and can be as tough as old boots. Such beef often weeps blood onto the plate when kept in a fridge overnight. Supermarkets are the main source of inferior, badly hung beef, whilst well-hung beef is mostly found at the traditional butcher or at farmers markets.

Another sign of quality meat is when it's marbled. That's where there are fine marbles of fat throughout the meat that helps to keep it juicy and give it extra flavour. Farmers and butchers have a big task ahead educating housewives to appreciate good beef. All that glisters is not gold and so it is with beef. Bright red, lean meat is often tasteless and tough while the dark brown, marbled steak is juicy and tender.

Anyway, that young couple have built up quite a reputation for their beef and their latest venture is to put a butcher's van on the road that goes round villages and remote houses selling their meat. Pensioners without a car really appreciate that service.

At one time there were hundreds of vans selling fish, groceries, bread and meat. They regularly called at farms and hamlets. It was a great service as well as the main source of local news and gossip. Three-quarters of those vans have disappeared in the last ten years in the face of stiff competition from supermarkets. Indeed that butcher's van is the first new one to be licensed in ten years. I can't help but admire their enterprise and wish them every success in the future.

14

Branching out

From time to time I find horseshoes on my land. They are often well-worn and rusty and must be old because it's more than fifty years since horses worked on my farm.

At one time they were everywhere. Powerful big Clydesdales drew carts or worked as pairs in front of the plough. Ponies pulled the trap that took the family to town, whilst magnificent thoroughbred riding horses carried the factor round the estate.

Horses could tell you a lot about the prosperity of a farm. My grandfather used to refer to farms by the number of working horses on it. 'Good place that. Eight pair o' horse, ye ken.' Obviously, better farms with big arable acreages supported greater numbers of working horses.

At one time the stables were the most important part of the farm steading. Well-cobbled, with high stalls made out of pitch pine, there was never a straw out of place. Horses, equipment and stable were all immaculate. On wooden pegs along one wall hung the collars and working harnesses.

I'm not old enough to have worked with horses, but my grandfather often talked about the long hours. Well before day-break the horses were fed, the stable was cleaned out and harnesses were fitted. After a long, weary day in the fields – which involved following a horse plough or set of harrows on foot – the horses had to be groomed and fed before the men could think of their own supper.

Despite the hard work, grandfather loved his gentle charges. There was always a gleam in his eye as he talked about them. Every farm was the same. Horsepower was the only source of power. All they needed was tender loving care and they would work without incurring garage bills or expensive spare parts. They even had the ability to breed their own replacements, unlike cars and tractors that have to be traded in and replaced at exorbitant rates.

Gradually, steam took over towards the end of the nineteenth century. Big steam engines stationed at either end of the field pulled ploughs to and fro by steel cables. Steam-driven threshing machines, mills, water pumps and the like became common on the larger farms. Mind you, horses were still needed to cart coal to fuel the mechanical devices that had replaced them.

Eventually modern tractors with pneumatic tyres and hydraulics won the day, and horses all but disappeared. Happily, they're on the increase again as everyone seems to want to own a horse or pony. That gives us farmers the chance to earn extra income by offering livery.

A pal of mine now runs a very lucrative equestrian business. For twenty years he milked eighty dairy cows, morning and night. Despite putting in long hours he found that he often lost money. Then passers-by spotted his wife's horse and asked if they could keep their horse on the farm. Being on the outskirts of Glasgow he soon had lots of horses without advertising. The more horses that grazed the fields the more new business he attracted; so eventually the dairy cows were sold and he now looks after fifty horses.

For about £60 a week he provides first-class board and lodgings for a horse. Those owners prepared to do the graft themselves, such as mucking out, only pay £30 for a stable with feed provided. He reckons he's never been so well off, although it's a totally different business. He now works with all kinds of folk at all hours of the day. That soon taught him that the customer is always right.

A building that used to hold thirty cattle is being converted into yet another five stables. Those five fee-paying cuddies will earn many times more than thirty dairy cows ever could. As well as renting stables and grazing he also makes money by selling hay for horse feed, but, despite his thriving equestrian business, he still keeps 160 beef cattle. Like most of us he is loath to part with his beasts and that baffles economists.

Diversification is the current buzzword but the idea has been around for a long time. The idea is to encourage farmers to become less dependent on agriculture for their income. That way, if prices drop for farm produce, farmers have another income to fall back on. All sorts of ventures have been tried.

Renting-out land for shooting and rivers and lochs for fishing were obviously successful with the right type of farm. Some found that coarse fish in lochs or ponds could be netted and sold to re-stock coarse fisheries whilst others reared crayfish. In the absence of heather, woodland, loch or river, clay-pigeon shoots were developed. Golf-driving ranges and golf courses met with success, but needed a lot of investment. Letting redundant farm buildings as workshops, or converting them to kennels or farm shops, is also a well-tried money earner.

Contracting with surplus machinery, sheep shearing, pregnancy scanning and the like are other farmer favourites. Not so successful were some of the exotic livestock ventures. For a time goats that produced luxury fibres such as mohair and cashmere were all the rage. Suddenly the bubble burst and those who had invested thousands in breeding stock found the animals became virtually worthless overnight. Others turned to llamas and alpacas as alternative sources of exotic fibres, with mixed results.

Farming deer for venison has had a steadier though unspectacular expansion. The attraction of venison is that it's delicious, healthy, fat-free meat. The main snag is that farmers have to market the venison themselves. Deer also need expensive high fences and specialist handling-pens.

Some adventurous types milk buffalo to make mozzarella cheese. Wild boar and bison have been introduced as gourmet meats, but these animals are dangerous and farmers need special licences to keep them. Anyway, I'm too portly to win the race to the gate if they decide to charge!

One craze that swept through the ranks of the innovators was ostrich farming. It's claimed ostrich meat is lean and tastes like beef. The skins make fine leather garments while their feathers can be used as fashion accessories. Despite costing a lot, enthusiasts reckon ostriches have a profitable future. That may or may not be so. Undoubtedly ostriches have the ability to drain the purse of the unwary. Apart from the significant expense of breeding stock, there are also high fences and shelters to be constructed. They can also be dangerous and have to be handled with care. While gentle handling will tame virtually every creature, there are always the stressful times to consider.

The thought of running after a bunch of young ostriches that have broken free makes me breathless just thinking about it. Having experienced the drudgery of plucking geese and turkeys, I can't imagine tackling one of those birds and the thought of tucking into an ostrich drumstick or boiled egg sounds daunting. Call me old-fashioned and set in my ways, but I think I'll stick to black cattle, sheep and free-range hens.

The best chance of success in diversification lies with tourism. City folk love the countryside and are fascinated by farms and farm animals. So farmhouse bed and breakfast, properly done, is as lucrative a new venture as you will find. There are also open farms where tourists can get close to farm animals or watch cows being milked from specially constructed observation platforms. One farmer I know has even opened a visitor centre where you can spend the day watching red kites.

Even I diversified seventeen years ago when I was fortunate to get the job of writing a weekly column for the *Sunday Post*. My only investment was a pen and writing pad and over the years the income has been greatly appreciated. Writing about day-to-day events on my farm has given me a lot of pleasure as well as extra income.

Others have developed holiday cottages and caravan sites and gone in for tourism in a big way. Another friend has done just that. His farm is in a beautiful part of Scotland. Tourists regularly pestered him to park a caravan overnight in his fields. That eventually developed into a major caravan site with all mod cons: adventure play-park, swimming pool, restaurant, pub and wet-weather recreation areas. On top of all that he organises hillwalking, pony trekking, mountain biking, canoeing and abseiling.

Nothing is too much hassle. He only wants to please and make sure everyone enjoys the holiday. From first thing in the morning till last thing at night he is at their beck and call: phoning to book theatre tickets, accessing weather forecasts, providing tourist information or fetching a doctor. He once told me that a field full of folk takes a lot more looking after than a field of cattle. I'll stick with livestock and continue to check them morning and night. What's more, they seldom disagree with me, complain or argue back!

Quite a few farmers have discovered that radio masts and wind turbines are lucrative new crops to plant on top of their highest, windswept hills. Our hills and glens make radio communications difficult. To improve their network, mobile phone companies and the emergency services need more transmitters. They range from a simple, fifteen-metre-high, public-broadcasting mast to monstrous forty-five-metre ones that can earn double or treble the rent. The money is not to be sniffed at; every transmitter fitted to the mast can bring in another thousand or two in rent every year. So the lucky farmer with a mast holding transmitters for BBC, Vodafone, Cellnet, Mercury, Orange, police and water authority is on to a nice little earner.

Wind turbines are also the words on many hill farmers' lips these days. Government has decided we need to generate some of our electricity from renewable resources. Not so long ago generators driven by a windmill were inefficient, but the scientists have refined such systems beyond all recognition. The first wind farms, erected in the north of England, had to be subsidised as the electricity generated was far too expensive, but the latest generation of wind turbines produce it at less than half the cost of those early models.

The west coast of Scotland is particularly suited to wind farms, as it's one of the breeziest parts of Europe. Developers erect the wind turbines on a suitable ridge and then pay the farmer an annual royalty, depending on the amount of electricity generated. In a windy year that royalty can run to tens of thousands of pounds.

The problem is getting planning permission for such developments as many folk find them unsightly and reckon they spoil our beautiful landscape. That may be so, but beauty is in the eye of the beholder.

I never liked large-scale conifer plantations but few others seemed to object when they were planted. As one who has witnessed trout disappear from our burns due to acid rain, I welcome every effort to clean up our environment. It seems to me that instead of lots of small wind farms scattered about the country we might be better to have a few large ones in discreet locations where forty or fifty windmills could work away without causing offence. Snag with that is the potential income from such development will end up in the hands of a few instead of being fairly shared out among many.

There are other sources of non-farm income that are often more hassle than they are worth. More than just farmers have the right to use their farmland. Everywhere you look, electricity cables and telephone lines criss-cross the countryside. Pylons, poles and their struts can be a real nuisance. At hay, silage or harvest time, they're often in the way of tractors and machinery. British Telecom and electricity companies recognise this and make small annual compensation and wayleave payments to the farmers affected.

Then there are all the underground services like gas and water mains, as well as underground ducts and buried cables.

In many instances the farmer has no choice in the matter, as services like water have a statutory right of access. Again compensation is paid for the disruption that installation brings, as well as the damage done to crops or underground field drains, fences and the like, but such payments don't fully cover the inconvenience to the farmer. Disturbed drainage systems take years to settle down again after having been dug up to make way for pipes and cables. The replaced earth continues to settle so that repaired sections of field drains sag out of alignment and block up with silt.

Biggest hassle is with railways, roads and motorways.

Rail tracks are bad news. Railway embankments are often overrun with rabbits that destroy adjoining crops. Fences are often badly maintained, so woe betide the sheep or cow that breaks out on to the railway line. Railway embankments, like road and motorway verges, are a haven for weeds and their seeds, leading to the contamination of adjacent fields.

Serious problems often arise when a new road cuts across the farm. The threat of a compulsory purchase order leaves the farmer with little say. How do you get farm vehicles and livestock safely across a busy new road? The answer is simple but expensive. You either build a bridge or tunnel beneath. Construction companies prefer tunnels, as they're cheaper to build. Snag is that they are often dark, badly drained and liable to fill with water. End result is that livestock are chary about using them.

As I said, there are systems to compensate farmers, but the compensation is often inadequate. It's also very difficult to extract payments from large companies or the Scottish Executive. There's invariably a never-ending legal process and, while the wrangling continues, affected farmers have to suffer the daily inconvenience of that motorway or bypass.

Until as recently as thirty-five years ago, most farms gave

employment to a fair number. Workers had to live on the farm to tend animals through the night and early in the morning. Those who didn't work with animals found it more convenient than relying on infrequent or non-existent public transport.

Bachelors working on larger farms often stayed in bothies, but most farm workers lived in tied cottages. Such cottages were offered rent-free and were regarded by farmers as a valuable perk. Farm workers may well have viewed their houses differently. Often cold, draughty and damp, those cramped little cottages were well below the standards of a council house. Many were without electricity and running water until the late fifties.

Eventually, economics forced farmers to lay off their workers and as a result the cottages also became redundant. After lying empty for a while, some were sold off as holiday or retirement homes. It wasn't uncommon to see two or three cottages being converted into one house, proving how cramped they had been.

Such sales helped reduce the farmer's overdraft at the bank, but others were simply asset-stripping. New owners of a farm would sell off cottages and small paddocks to reduce the purchase price. Prudent farmers did their cottages up and rented them out as holiday homes, giving farm incomes a badly needed boost.

Holidaymakers, and the retired folks or commuters who now live in these cottages, are very different from farm workers in many ways.

They don't necessarily understand farming, nor are they as tolerant of our country smells, dusts and noises. Not being employed by the farmer, they aren't afraid to complain about such nuisances. Folk seeking injunctions against farmers to restrain them from various farming activities are now fairly common. The new occupants of those cottages may well consider themselves country dwellers but, when they come up against country smells like slurry, dust from cultivations or the noise of a grain dryer, it's a different story!

15

That's progress for you

As in every industry, farming is being bombarded with new technology, and sometimes it can do wonderful things. For instance, a friend lost all his sheep in a cull during the foot-and-mouth epidemic when a nearby farm became infected. Luckily, his beef cattle were spared.

That flock was no ordinary one as my friend had some of the finest sheep in Scotland. His speciality is Bluefaced Leicesters and he has won nearly every championship and trophy in Scotland with them. Bluefaces are a low-ground breed that is crossed with hill ewes like Blackfaces or Swaledales to produce mules. Mule ewes are excellent mothers and produce lots of twin lambs that grow big and meaty when mated with bigger breeds like the Suffolk or Texel.

As well as selling high-priced, pedigree Bluefaced tups, my friend also breeds mule-ewe lambs from his large flock of Blackface ewes and regularly gets top price at the autumn sales. So he was devastated when they were slaughtered. He reckoned that a couple of pedigree lambs born that spring were crackers and would have won many shows that summer.

Fortunately his best show-ewe was getting old so he had sent her away the previous autumn for embryo flushing. That's where vets use fertility drugs to increase the number of embryos ewes produce. The idea was to store those embryos frozen and implant them into other ewes the following autumn in the hope of breeding more daughters. As a result, that prize ewe escaped the cull and was kept safe at the clinic, which was well away from the infected area. The vets had already flushed and stored eight embryos and went on to produce many more.

My friend also had the good fortune to have stored a large quantity of frozen semen from a champion tup that had died a few years earlier. Thanks to modern technology he saved those precious bloodlines and was able to rebuild his pedigree flock. Many of his test-tube lambs went on to win shows and breed more show winners.

Restrictions on the movement of livestock due to foot-and-mouth led to markets being banned although farmers had to continue buying cattle to fatten otherwise their grass would have been wasted. Fortunately, as the outbreak came under control the rules were relaxed so that livestock could be moved from farm to farm under special licences. Snag was, finding a fair way to value the animals. At the end of the day you can't beat the auction system to determine true values. So auctioneers came up with a system of selling livestock by video. Aberdeen and Northern Marts pioneered the system and made videos of the batches of cattle for sale. Buyers were then able to view the videos before the sale, either at the market or at home through the internet. On the day of the sale, videos of the cattle were displayed on a big screen in the market and there was also a catalogue that accurately described them.

The system worked reasonably well and was much better for the cattle. Beasts didn't have to suffer the stress of being transported to a market and were simply taken directly from one farm to another. Apart from eliminating the risk of spreading foot-and-mouth the system also prevented the spread of other diseases. Farmers could still enjoy the social side of going to the market and sellers knew they were getting a fair price. Better still, market staff didn't have to clean up the sale yard afterwards. You would be amazed at the amount of muck several hundred cattle leave behind after a day at market.

Some speculated that old-style markets would be a thing of the past and that video sales were here to stay. They hadn't reckoned on the canny nature of farmers. Once foot-and-mouth was over, and things returned to normality, they turned their backs on those new-fangled video auctions.

Over the millennia man has proved to be a master breeder. We have selectively bred wild plants and animals to suit our own needs. Look at any garden and see the range of flowers that have been bred. Roses of different shapes, colours and fragrances prove that breeders have long been at work creating undreamt of beauty. Horticultural shows have displays of vegetables from the ordinary to the unbelievably massive. Foresters have also seen the benefit of breeding programmes to improve profits. They have selected the fastest-growing, tallest trees to breed the next generation. More recently they are selecting trees that grow straight and yield stronger timber for the construction industry. It's the same on the farm. Modern varieties of grain are bred for specific uses like baking bread or for distilling into our favourite tipple.

Over the years breeders have developed varieties with important traits like early ripening, heavy yield or simply the ability to withstand torrential rain without being flattened. It's the same with our farm animals. Horses now range in size from the tiny Shetland pony to the mighty Clydesdale. Cattle come in all shapes and sizes from the petite, brown, Jersey dairy cow famous for giving rich, creamy milk, to the massive Aberdeen Angus with its tender steaks. Amazingly, we're still not content and strive ever harder to breed perfection.

Scientists around the world are now using genetic engineering to bring about even more sophisticated developments, like crops resistant to disease or animals that can supply us with organs for transplants. The latest advance provides the potential to breed sheep with coloured wool. Researchers in California have managed to make white mice grow green hair. It's part of a programme that's trying to reverse the greying of human hair and looking for a possible cure for baldness, but the results could also be applied to sheep. A gene has been taken from a jellyfish and inserted into the hair follicles of mice. That gene makes a protein that glows green when illuminated with blue light. Scientists reckon they can also transfer that gene to sheep and breed a strain with coloured wool. That could end the need for dyeing wool and help shepherds spot brightly coloured, stray sheep on their hillsides. We may be on the brink of a breakthrough and Scotland's tourist board could have a new attraction as the first flocks of tartan sheep appear on our hills!

One of the main reasons for the low price of wool is competition from other fibres. Apart from sheep with their woolly fleeces, most animals that graze mountains have a shaggy coat of long hair to keep the rain out, as well as an undercoat of fine hair next to their bodies that keeps them warm. Those fine hairs are extracted to make cashmere and mohair. Main suppliers are countries like Mongolia, China and South Africa, which run large herds of fibre-producing goats. There are also fine fibres from South American llamas, alpacas and guanacos. Often overlooked is silk produced from Chinese silkworms and don't forget fibres from plants like cotton, or linen produced from flax.

There are also synthetic fibres like nylon, rayon and the polyesters. Scientists keep inventing new ones. The Japanese once developed a fibre from protein found in milk, but fortunately for sheep farmers it was too expensive. Recently the Chinese have developed the world's first cashmere-like garment using a fibre made with protein extracted from soya beans; good news for arable farmers who can grow soya. China produces 8,000 tons of cashmere each year, about 80 per cent of the world's output. Sadly, all those cashmere goats are damaging China's hills and valleys. As well as munching everything that grows, goats also dig up roots for food, so a goat does more damage than twenty sheep.

The Chinese believe their new soya development will lead to a reduction in goat numbers and protect their environment. It won't do anything to help my wool prices, but luckily it will take years to develop a cloth made from the new fibre. Until then I'll keep wearing warm woollies!

One of the biggest changes in farming has been brought about by the development of all-terrain-vehicles (ATVs), those little four-wheel motorbikes. They are now so popular they could be described as indispensable. Lightweight and fitted with wide balloon tyres, they hardly leave a wheel mark on the wettest land. That makes them invaluable to arable farmers.

Slugs and fungi often attack autumn-sown crops when the weather is wet. Driving on sodden fields with a tractor and conventional machinery can make an awful mess, but an ATV can tow a small hopper and spinner that will spread slug pellets.

Alternatively, there are small, low-volume sprayers that fit on the back of ATVs. It's amazing the number of different attachments that are available.

Livestock farmers, in particular, make great use of ATVs and the four-wheel-drive versions can go almost anywhere. They'll climb all but the steepest of braes and cross the roughest of terrain, saving the shepherd's legs. As a shepherd once said to me, 'It'll go places I'm not prepared to go!' Even sheepdogs learn to cadge a lift by sitting pillion behind their master.

Lightweight aluminium trailers can be towed to carry feeding to the highest peaks or transport sick sheep. ATVs allow shepherds to look after twice the number of sheep. In fact, it's not just shepherds that run the risk of being made redundant by those noisy workhorses. Sheepdogs are also under threat!

Many young farmers find that they can round up sheep and cattle just as easily with an ATV as with a dog. Some young farmers clubs even hold trials like the ones in the popular television programme, *One Man and His Dog*. Only difference is, there isn't a dog in sight, just a bunch of lads on ATVs. I have even heard that some reckon that they could beat a sheepdog. That may well be so, but, at around £5,000, they don't come cheap. Like all machines, they don't always start in the morning, and can break down. Eventually they rust away to worthless scrap and, unlike a pair of working collies, they can't breed a replacement or show affection.

Enough said!

Double glazing was one modern improvement that didn't work as well as I had hoped. Newly born calves run the risk of being trampled in the main cowshed so, once a cow has started to calve, she should be moved to a calving pen. Luckily I'm a light sleeper and can usually hear the distinctive noise of a calving cow. Then it's simply a matter of getting out of bed for twenty minutes and moving the cow to a straw-bedded pen, thus allowing her to calve in comfort.

Unfortunately we fitted the farmhouse with double glazing in the summer of 2000. My wife reckoned that was a big improvement on the old draughty windows that also let in dust.

Snag was, I could no longer hear my cattle through the double-glazed windows and I had to get up through the night to check them. There's nothing worse than needlessly losing precious sleep so I came up with a solution to the problem, although it was one that didn't please my wife. I left our bedroom windows open so I could hear the cows as I used to!

Then I found a better solution when my wife and I spent a weekend with farming friends in Northern Ireland. At night we watched a new television programme that revolved round a cast of fifty pregnant ewes, in which the main storyline centred on several of them lambing in the middle of the night. You see, my friend had recently installed two closed-circuit television cameras in his lambing shed. That way he was able constantly to check his flock from the comfort of his fireside or bedroom. It was a fantastic idea and well worth the £400 that it cost him to install. Instead of getting up through the night, donning boiler suit, heavy coat and wellies to trudge across a cold, wet and windy yard, he simply rolls over in bed, presses a switch and looks at the screen. Often as not, nothing is happening; so he simply goes back to sleep. Under the old system it would have taken twenty minutes, woken his wife and taken far longer to get back to sleep.

Truth to tell, you can see far more with closed-circuit television. Opening the lambing-shed door and walking along the feed passage disturbs sheep. As they bunch up you then have to let them calm down before slowly walking past to check that they're OK.

Closed-circuit television doesn't disturb the animals. It's easy to spot an agitated ewe about to lamb among the rest that are peacefully lying down. Spying on undisturbed sheep lets you observe a lot more. You easily spot ewes nudging one of her lambs away as she starts to reject it. Lambs that you weren't sure had learned to suckle properly can be clearly seen feeding themselves. Yes, closed-circuit television is a breakthrough in caring for stock, but not a lot of help to my flock. I keep about ten times more sheep than my Irish friend and they all lamb outdoors.

Still, closed-circuit televisions could have advantages in my cattle sheds. It will probably cost a little short of a thousand pounds, but I'll easily recover that by saving a couple of calves and having a good night's sleep. Better still, my wife won't moan as much about my cold feet when I get back into bed, and it will make more interesting viewing than what's on the box nowadays!

16

Mud, glorious mud

One of the problems with being married to a house-proud wife is her insistence that I keep everything around the farmhouse spick and span. In common with a lot of farmers, I am not the most dedicated of gardeners. Farmers are used to growing crops on a grand scale. Digging a garden with a spade seems complicated and tedious compared with ploughing a field with a tractor and reversible plough.

Weeding a vegetable plot doesn't begin to compare with spraying a field with herbicides using a tractor-mounted sprayer, whilst cutting grass on a lawn is a nightmare compared to mowing a field for hay or silage. Yes, farmers are used to tending their land with tractors and machinery and don't take readily to the laborious, peasant-style conditions in a farmhouse garden.

Don't underestimate the scale of the problem. Many farmhouses have huge gardens, unlike those in towns with their petite dimensions. Farmhouse gardens were designed on a grand scale to provide vegetables all year round as well as fruit for jam. The need for thrift was driven by the number of hungry mouths to feed in a typical farmhouse of yesteryear.

In addition to the farmer's family there were often maids or single lads living either in the farmhouse or in bothies. On top of that there might well have been other farm staff to feed during the day, particularly at hay time, harvest or at clippings. When times were hard, particularly during the depression of the thirties, it was essential that money was not spent needlessly on groceries that could be grown at home.

All those hungry mouths could be relied on to help in the garden. A press-ganged squad of maids, farm workers and youngsters made light of most routine tasks in a farmhouse garden. Sadly, few of us can afford to hire farm staff nowadays. Those that do would never dream of asking hard worked and skilled staff to weed a vegetable patch, or cut the lawn. End result is often a neglected garden.

In my early years I occasionally helped my wife – who is a very dedicated and enthusiastic gardener – to keep the garden tidy. I soon discovered that it took a lot of hard work to produce very indifferent-looking vegetables that were either diseased, or had been partially chewed or burrowed through by just about everything you can think of, ranging from tiny carrot flies to slugs, caterpillars, rabbits and the like.

Hens wander in and make dust baths in flowerbeds, discarding seedlings in the process. Hens, pheasants and pigeons love peas, lettuce, cabbages and young cauliflowers. Sheep like to jump the dyke or crawl through the fence in search of tasty, crisp vegetables. Oh, and I nearly forgot to mention cattle. Several times in my gardening career, I have watched forlornly as a bunch of cows disappeared through the garden gate that had been left open. They hardly had time to eat a thing, but flattened everything in sight with their lumbering hooves.

Biggest problem with farm gardens is the weeds. Unlike townies who buy sterilised horse manure, compost or fertilisers from garden centres, we thrifty farmers have tons of free muck. The drawback is that farm muck isn't always properly composted when it's needed, and can be full of weed seeds. Slurry – that is, cattle muck without straw – seems to encourage chickweed.

I remember once at a market discussing with a farming friend whether slurry or muck was best. Another farmer butted in, 'If you want a garden kept free of weeds, it's not muck or slurry you should put on, it's four inches of concrete!' I eventually came to the conclusion that it would be far simpler to lay the vegetable plots down to grass and buy vegetables from the supermarket.

Had I been married to someone else I might have been allowed to implement the more radical plan of removing the surrounding dyke and incorporating the garden into the adjacent field. My wife then set about planting herbaceous borders that she tends with a degree of obsession that frightens me! They may look like flowers to her, but if I found them growing in a field I would consider them weeds. Anyway, gardening is her hobby and she gets a lot of pleasure from it.

The most frightening aspect of my farmhouse garden is the capacity for the lawns to grow grass. Even in the most difficult growing season the lawn grass flourishes, while my cattle and sheep can hardly find a bite to eat in the fields. That's all the more remarkable when you remember that the fields are regularly fertilised whilst the lawns never get an ounce. Perhaps it's time we did some research on this strange phenomenon.

Growing Christmas trees is the only thing I can think of that is more complicated than gardening. It's a specialist crop that takes six years from planting before the fastest-growing ones are ready for sale. After that first thinning, the rest are sold over the following three or four years before the field is replanted. Farmers get about £12 for a six-foot-tall tree although shopkeepers and traders charge between £25 and £35.

With more than three thousand trees per acre, a twenty-acre field is worth about £2 million. No wonder farmers get nervous in the run-up to Christmas. Fortunately, nobody wants Christmas trees in November and not even Del Boy could sell one on Boxing Day, so the high security only lasts for about three weeks.

Christmas trees are more fickle than sheep and everything to do with them costs money. Fields have to be chosen carefully as new growth is very prone to frost damage, so a north-facing field is preferred. That way growth starts later in spring when there's less chance of frost. The field should also be on a slope so if there is a frost the cold air will quickly move down to the bottom of the hill.

Wildlife is a major problem. Rabbits and hares nibble the young trees, while deer love to rub the velvet off their antlers against them. So expensive fencing is needed to keep those unwelcome intruders out.

Birds naturally love to perch on top of Christmas trees. Unfortunately, when they fly off they break the delicate top bud. Without that bud, Christmas trees grow bushy rather than conical shaped. So special perches have to be erected to encourage birds to rest elsewhere.

Fertilisers have to be applied by hand to the base of each tree because spreading it mechanically by tractor would damage them. Another reason for carefully placing it by hand is that when chemical fertilisers land on the trees, they scorch the new growth.

Some farmers use a breed of sheep called Shropshires to graze down the weeds. Shropshires are chosen because they hate the taste of Christmas trees and won't nibble them. Finally, after about three years of growth, you have to start thinning-out bad trees and pruning the rest to get them in perfect shape. No wonder Christmas trees are expensive and gardening doesn't seem so bad compared to growing them!

My wife's hankering for a spotlessly clean farmhouse eventually led me to solve another long-standing problem: that of mud, glorious mud. Our farm steading sits on a hillside. The long, steep, farm road continues up through the steading to the hill. When it rains heavily, water runs off the fields and rushes down the road into the yard. We have built run-offs that are supposed to guide the water into a roadside ditch. Despite being meticulously maintained, they very quickly silt up in a flood. The worst time is when snow thaws quickly as a result of rain. Lumps of half-thawed snow block the outlets to the ditch and water floods down the road. All that floodwater eventually ended up at the back door of the farmhouse as the drains in the yard silted over. After a flood the mess was horrendous

The previous tenant had tarred part of the yard at the back door. No matter how thoroughly you swept, the silt stubbornly stuck to the chips in the tar. Why my predecessor tarred that part of the yard is beyond me as a cobbled yard is always far cleaner looking. Dirt that falls between cobbles is never noticed. Every year I scrape and sweep the cobbled part of the yard and it's amazing the barrow loads of dirt that come off an apparently clean yard. Sadly, even if I could have found someone skilled enough, it would have been too expensive to lay new cobbles in place of that dirty, tarred area.

For years my wife was tormented with muddy boots and shoes at the back porch until her patience eventually snapped and I was ordered to do something about the mess. A quick phone call summoned a local builder and, after much discussion, we decided to concrete the offending part of the yard and to rearrange the drainage. We reckoned that better-sited drains could prevent the flooding and if they failed and we had a mess, we could easily sweep it up and then hose the concrete clean.

Nothing is ever as easy as it seems. For weeks, lorry loads of gravel and sand lay tipped in the yard. There were also pallets stacked with cement bags and covered with tarpaulins. Then there was the digger that removed the old tar and laid the new drains.

All the mess drove my wife frantic. I'm sure she had second thoughts as she constantly wiped the dust that somehow got through closed windows. Eventually, we laid the new concrete and I must say it looks marvellous.

Next morning, when I went to see if it had set properly there was a trail of footprints across it. During the night one of the cats had walked across in search of food. Those tracks will be there forever, but it was the last time that footprints were found at the back door!

17

Sanquhar show

Early spring till autumn is the season for agricultural shows, ranging from the mighty, four-day Royal Highland to local, one-day events. They are a grand chance for farmers to have a day off, meet old acquaintances and catch up on the crack.

Bigger shows display all manner of farm equipment, seeds, feedstuffs and the like. There are also technical stands and displays explaining all the latest techniques. Elsewhere you will find craft tents and competitions with all the usual classes for baking, needlework, flower arranging and crook making. Above all else, there is the livestock section with its poultry, cattle, sheep, goats and horses.

An unbelievable amount of time and effort goes into preparing animals for a show. Over the winter months show prospects are fed secret rations that grow them into sleek creatures of perfection. They're carefully washed and groomed then trained to walk in such a way that they catch the judge's eye. It's all passed down through generations of farmers.

Watching the judging is always thoroughly enjoyable, as it is not only the animals that are being judged. Hundreds of experienced farmers are judging the judges as well! Later, in the beer tent, the judge's failings will be debated as fiercely as the animals on show. While fathers put the world to rights round the judging rings or in the beer tent, mothers take the children for a shot on a go-kart, bouncy castle or whatever. Towards the end of the afternoon the whole family gathers round the main ring to watch either the grand parade of show winners or a gymkhana.

Sadly, local shows are in danger of dying out because of lack of finance. They cost a fortune to run. There are all the tents, prize money and ring events to be paid for. Insurance is another escalating cost. Above all else it takes a lot of hard work by a dedicated committee with ever fewer volunteers. Then, when a wet summer comes along, the event may be cancelled or run up a massive loss through poor attendance. It will be a sad day if local shows disappear from the farming calendar, so I urge everyone to support them. They're a good family day out and a grand chance for townsfolk to learn about the country way of life.

In 1998 my local town of Sanquhar celebrated the four-hundredth anniversary of it being granted a royal charter by King James V1. It was a grand year full of different activities and, not to be outdone, the farming community decided to reconstitute the Sanquhar Farmers Society, which was originally founded in 1842. Its sole objective was to put on an agricultural show in Sanquhar, an annual event that had been enjoyed for ninety-nine years until it lapsed in 1932.

The earliest written record of the show is found in a local newspaper, the *Dumfries and Galloway Standard* of 23 July 1848. It reported that: 'The annual competition for the premiums offered by this society, took place at Sanquhar on 18th inst, when the stock was as numerous as in former years. The recent improvement in the quality of the stock was the subject of general remark, proving the advantage of the society to the district and the success which has attended its operations.' There was a comprehensive show of Ayrshires, sheep and pigs that year.

Farming profitability started to rise towards the end of the first half of the nineteenth century and the Sanquhar Farmers Society was probably founded to improve farming methods in order to take advantage of this period of relative prosperity.

Cheese more than doubled in price due to an upturn in demand from the colonies, especially Australia, where the gold fields were being developed. Another reason was the rise of activity in the iron and associated industries that followed from the development of a rail network. Next, it was the turn of the livestock farmers. In 1863, as a consequence of the American Civil War and the scarcity of cotton, wool prices soared. Hill lambs also rose in price and had nearly doubled in value by 1872.

Sanquhar held three Great Fairs as well as other smaller events. Candlemass, or the Herd's Fair, was probably the biggest event. It was a hiring fair and after they had made their bargain with the farmers for another term of employment the shepherds held a great celebration in the town.

The November Fair traded principally in vegetables and was known as the Onion Fair. Prime onions were sold in French fashion, braided on to straw strings, whilst the poorer quality was sold by weight. At these fairs, the sides of the main street were occupied with small stalls and booths. The south side of the street was reserved for the sale of boots, shoes and slippers. The evening was generally brought to a close by the dancing of penny reels at the council house.

The Sanquhar Wool Fair was held in July and was the great lamb and wool market of the year and, as it followed immediately after that of Inverness, it regulated the price of wool for the southern half of Scotland. Many dealers and flock masters attended, all interested to know the latest prices.

A report of the 1875 fair records that: 'prices ranged about the last year's rates, buyers declaring it was then purchased too dear. The recent heavy failure of the tweed trade had also an adverse influence on the wool market. Woolbrokers, commission agents and dealers in dipping materials were present in great force, and there was fair demand for Blackface wool. Cheviot was not so much enquired for and few sales in white wool were effected.'

Trade was brisk in those days: £100,000-worth of sales could be realised at a single fair in the 1870s. As the wool fair was held so early in the season it was known as a 'character market': that is, the stock was not shown, but bought and sold on the reputation of the farmer. Another local paper, of 25 June 1833, advertised that a 'tup shew' was to be incorporated in the July Wool Fair and was probably how the show originally started.

So we had a lot to live up to when, in the winter of 1997/98, a committee of twenty-five enthusiastic amateurs met to begin planning every detail of the one-off show. We reckoned it would cost £14,000 to stage, and that the money had to be raised from at least a hundred different sponsors.

Sub-committees were formed with a bewildering variety of remits. One to promote a horse show and gymkhana, others to organize a whole host of activities: classes for sheep, cattle and Clydesdales, a fleece competition, an industrial tent with craft workers, and classes for walking sticks, baking, preserves, vegetables, pot plants and handicrafts. Other sub-committees planned main ring events such as silver and pipe bands, country dancers, sheepdog and Clydesdale horse displays, tug-o'-war, a parade of livestock and presentation of prizes. There were also side events such as a vintage-machinery display, mini-quad bikes for children, face painting and bouncy castles.

Forty different firms were encouraged to set up trade stands displaying machinery and services. Water had to be piped onto the show site, while portable toilets, generators, security lights and hundreds of other forgotten items needed to be hired. A marquee for a dinner dance for two hundred had to be erected and laid out, as well as catering tents and a hamburger stall. Rosettes had to be made, medals and trophies designed. Just before the show, hundreds of gates were hired or borrowed to set up the cattle and sheep pens. Car parking had to be planned, and hundreds of white posts driven into the ground with rope attached to mark out the various rings.

During all this mayhem it never once stopped raining!

On the morning of the show we woke up to a sodden show field and constant rain. All vehicles – such as livestock lorries, ice-cream vans and exhibitors trailers – had to be towed onto the field with tractors, but we soldiered on and by mid-morning the sun was out. That dried out the field and drew a crowd of 5,500.

End result? A brilliant success that was a tribute to a hard-working committee that mucked in but never gave in. There was even a small profit for charity.

18

A load of rubbish

Every Easter the countryside yields the first crop of the year and there are always bumper pickings. Unfortunately the bounteous crop that farmers have to gather is the litter left behind after an Easter weekend. Not that Easter is particularly special for litter; rather it is the first popular outing in the year when visiting folks can make a mess.

Cans, wrappers and bottles appear all-year round thanks to motorists and their passengers snacking on the move and then casting their rubbish out of the car windows. Look at the vast quantities of unsightly rubbish lying on roadside verges, or polythene waving gaily in the wind as it hangs from trees, hedges or fences. To hell with the quaint custom of taking your rubbish home with you!

Mothers leave piles of disposable nappies underneath hedges at lay-bys. Garden rubbish is neatly stuffed into polythene bags and then thrown into roadside woods. All manner of tradesmen economically dispose of unwanted rubble, kitchen and bathroom fittings in any discreet nook or cranny rather than hand over their hard-earned cash at a landfill site.

Messiest of all are those campers or picnickers who enjoy lighting a campfire that is invariably fuelled by handy fence-posts, wooden rails or even field gates. After a sunny weekend, our visitors leave our hills and glens looking like the slums to which they return. I can only assume that such people live happily in slums because decent folk wouldn't make such a mess of our beautiful countryside.

Television adverts warning 'dumb dumpers' that they will be prosecuted are ineffective because the police can't cope with a situation that's spiralling out of control. Matters could get worse now that the Scottish Parliament has passed legislation granting everyone the right to roam. Much has been made of responsible access, but not all of our visitors will act responsibly. In fact, the highly publicised right to roam could well attract a yob element to the countryside that will enjoy vandalising it just as much as they do their home environment. Such yobs will regard the right to roam as a citizen's charter to block farm access by parking where they please, to flatten crops, poke about farm buildings and leave gates open.

Farmers may reason, cajole, plead or remonstrate till they are red in the face, but that is unlikely to change the attitude or behaviour of yobs with a chip on their shoulder and hell-bent on confrontation. I hope I am proved wrong and that people exercise their new right to roam with care and consideration, but I have my doubts.

Proof that a fair number of folk don't give a damn about the countryside can be seen in the urban fringes. Such farmland is virtually impossible to farm conventionally and is often abandoned. I suspect that in the future there will be slightly less litter at the side of the quiet country roads around my farm as it gets spread more evenly across the whole countryside.

Only a few farms have the luxury of getting rubbish uplifted from their doorstep. Despite paying council tax like everyone else, most of us have the inconvenience of taking our wheelie bins and rubbish bags to the main road. That can lead to problems when animals or birds tear open the bags for the titbits inside, allowing the rubbish to scatter in the wind.

At one time every farm had its own coup, often discreetly hidden in a nearby wood. Our own farm coup, unused for years, became a treasure trove for my children when they were young. They often brought home unusually shaped antique bottles to add to their collection.

Most farmers stopped using farm coups when councils agreed to uplift domestic rubbish. My own council has the distinction of being bottom of the household-waste recycling league in Scotland. Council bosses in their wisdom prefer to destroy the environment by continuing to bury our refuse in massive landfill sites.

They aren't alone in not cleaning up their act. The United Kingdom was threatened with infraction procedures from the European Union. Put simply, the EU took legal action against this country for not applying relevant legislation to farm waste. It is estimated that if all agricultural waste is sent to landfill it will cost British farmers £45 million a year.

Biggest problem is with plastic film, which is used in large quantities to cover silage. This is because the burning of plastic and tyres on farm has been stopped. It's also now illegal to bury such rubbish on farms. That's a big problem. British farmers produce a staggering half a million tons of waste plastic every year. The problem with plastic film is that it doubles in weight when it becomes dirty. Plastic sheets may well come neatly wrapped on a short cardboard tube, or folded in a polythene bag, but once they've been used they seem to increase in volume a hundred-fold.

It's their awkward bulk and dirtiness that makes farm plastics unattractive to the manufacturers who specialise in recycling. They are not only expensive to transport, but also have to be thoroughly washed before processing. That's a pity, because proper disposal of plastics will be a compliance issue for the new Single Farm Payment subsidy that will be overseen the by zealous officials who make farm visits.

The growing cost of farm-waste disposal is a concern. I spent

£310 last year hiring two skips to get rid of old fencing wire, scrap corrugated iron and other awkward rubbish. On top of that I have to pay the Scottish Environmental Protection Agency for a permit, simply for the dubious privilege of spreading sheep dip on my own land. Then there are all the abattoir costs of offal disposal that are passed back to farmers in the form of reduced prices.

My biggest gripe is the needless expense of disposing of dead animals now that it's illegal to bury them on the farm. That cost will become very hefty indeed when the current financial support for the scheme is withdrawn, yet it's still perfectly all right to bury humans.

Although litter and refuse is a growing problem for farmers it's amazing how useful some rubbish can be. Empty ice-cream containers or margarine tubs are handy for storing nuts, bolts and nails in the workshop. Plastic washing-up bottles make grand oil pourers; their plastic nozzle is just the right size to fit into the lubricating hole of a shearing handpiece. Used tyres make ideal weights to hold down plastic sheeting over a silage pit.

Lambing time is when we seem to find a use for all kinds of rubbish. Oil drums with one end removed make great incubators. Line one with straw, suspend a 250-watt bulb over it and you have got a sure method of reviving lambs suffering from hypothermia. When they have recovered they can be fed milk from an empty whisky bottle because rubber teats fit them perfectly. Sitting by the fire in the evening it's a pleasure to empty the bottle knowing it could save a lamb's life!

Propionic acid for preserving moist grain comes in five-gallon plastic containers and sawing the bottom half off makes an ideal individual trough for sheep. My best ploy has helped to save hundreds of lambs from perishing over the years. I was in town one day and saw a skip loaded with what looked like flimsy, broken pallets. Closer examination revealed that they were packing-case sides made out of pallet-type wood. At about three-feet high and six-feet long, I reckoned they would make ideal individual pens for ewes and lambs.

Not one to miss a chance, I asked the owner of the skip-hire firm if I could have those sides. He was equally sharp and asked forty pence apiece for them. It went against the grain to pay hard cash for something I knew was going to be dumped, but that's business and anyway I reckoned they were still cheap. Back home we put them together and nailed on a few extra spars of wood to prevent lambs from crawling through. After a hard night's hammering we now have sixty individual pens in the lambing shed.

As soon as a ewe lambs in cold, wet weather we bring her into the shed and keep her individually penned with her lambs for a couple of days. Once they are two days old most lambs can cope with all but the most inclement weather. Beauty of keeping the lambs separate with their mothers is that they don't get mixed up and run the risk of being rejected by their ewe.

It's great how one man's rubbish is another man's gold!

19

Better safe than sorry

Health and Safety inspectors regularly make unannounced visits to farms and they can take tough and immediate action against anyone who breaks the rules. The inspectors are especially interested in machinery being cleaned while motors are still running or where children are being placed at risk; and so they should, because we all need to be reminded how dangerous farms can be.

It's all too easy to save time by not replacing a guard after a repair, but those precious minutes saved in the mad rush to finish a job could cost a life. Coats, scarves and trousers can get caught up in drive shafts, chains or gears; elsewhere, there are risks from injury by livestock, chemicals or poisonous fumes.

Most at risk are children. Kids find farms exciting places to play. There are stacks of bales to climb or to build houses in, workshops and chemical stores to explore, and sheds full of combines, tractors and four-wheel bikes to climb on for a pretend drive.

Apart from keeping such things under lock and key there is also the never-ending task of supervising energetic children. That was no problem in the days when there were lots of workers on a farm. When I was a lad there was always an 'auld tell-tale' nearby to tell my father what I was up to, or give me a clip round the ear. Nowadays most farms don't employ anyone and the farmer is busy from morning to night.

Worse, as farm incomes have declined, wives have had to go

out to work, so dad often has to watch the kids after school. Easier said than done when crops have to be harvested, cows milked or livestock fed. Older kids enjoy helping father, but it's the wee ones who are the problem. A split second is all that it takes for one to be run over by a tractor. No matter how busy you are, you must always remember securely to cover the sheep dipper, slurry sumps, grain pits and the like to prevent a child falling in and drowning. Similarly, you must never lean heavy gates against walls because they could fall and crush a wee one; above all else, you must have eyes in the back of your head.

Accident statistics often include children who fell through asbestos roofs, drank poisonous chemicals, drowned in slurry lagoons or were maimed by tractors and machinery. Almost worse than the accident must be the lifetime of remorse, because the farmer will also remember that carelessness led to an unnecessary accident.

Nowadays we continually carry out risk assessments. Instead of wearing a facemask to protect lungs from the dust in a barn, the idea is to avoid creating the dust in the first place. Either redesign the mill or ventilate the barn differently. Similar thinking is to prevent the development of dangerous mould on grain or hay rather than trying to avoid breathing in the mould spores that cause farmer's lung.

So risk assessments make farmers sit down and think about how they work, with safety as the priority. Then having thought about and perhaps solved some of the problems, they write down everything in a report. Therein lies the snag, because most farmers hate office work. They would rather be outside working than stuck at a desk writing a report, but it's not as bad as it sounds.

Apart from saving lives or preventing illness, a safety assessment can improve the farming operation. After all, mouldy grain or straw isn't particularly valuable or nutritious. Many would think farming is a healthy way of life: stress-free, hard work in invigorating fresh country air. That may be the case if you are not dependent on the farm for a living and only visit it for a break, but the pressure of working against the weather, bankers and seasonal deadlines all take their toll.

Next to doctors, we farmers are the second-most-likely occupational group to commit suicide. Apart from the pressures there is also the environment. Arthritis and rheumatism are brought on by a lifetime working in the cold and wet; back troubles through lifting awkward, heavy weights.

Elsewhere we have been made ill spraying insecticides, or using sheep dips. Admittedly, we now wear protective clothing, but for many of us it's too late. In 2005, a report commissioned by the European Union strengthened the suspicion that pesticides can cause Parkinson's disease. The Goeparkinson study that was published in the *New Scientist* involved nearly three thousand people. Headed by Anthony Seaton of Aberdeen University, it concluded that farmers were 43 per cent more likely to develop Parkinson's because of their high exposure to pesticides. That should come as no surprise; after all, if pesticides kill bugs, it's reasonable to expect them to cause ill health in those who work with them.

Sheep dips are a classic example of the dangers of pesticides. Sheep are constantly under attack from all sorts of bugs and creepy crawlies. In the spring there are ticks that attach themselves to the sheep to feed on blood that can pass on infections like tick fever. Summertime is when greenbottle flies lay their eggs on soiled parts of the fleece. They hatch out into maggots that can literally eat a sheep alive. Autumn and winter sees sheep scab mites and lice attacking the sheep. The best defence is to dip each sheep in a solution of pesticide known as sheep dip. Those insecticides kill off the bugs and prevent re-infection, but dipping isn't as straightforward as it sounds.

Sheep may appear stupid to the novice, but they always remember the smell of dip. That reminds them that they are about to be plunged into a cold bath. Most sheep stubbornly refuse to go near a dipper and have to be manhandled. It's a sweaty job made worse by cunning sheep. Many are the times I have slipped in the mucky pens and fallen into the dipper myself. So I know why sheep aren't keen on being dipped!

I stopped routine dipping some years ago and now only dip when there is a need. Older dips like DDT were highly toxic and entered the food chain of birds of prey like kestrels and sparrowhawks, making them sterile. That's why DDT was banned. Mind you, it wasn't just farmers who used insecticides; they are also used by thousands of gardeners to protect flowers and vegetables. A new range of dips was developed based on organophosphates that were originally used in nerve gases in the trenches. They are very effective in protecting sheep, but there's now evidence they cause ill health in humans. Hundreds of sheep farmers and shepherds have complained of being unwell after using them and a fair number now suffer from long-term ill health.

As a result, there are strict rules to control dips. Operators have to be trained and pass an exam before gaining a certificate that allows them to use them. Specialist clothing, gloves and facemasks that make the farmer look like a spaceman have to be worn. Even the latest dips, containing synthetic pyrethroids, have question marks over their safety. So I have done away with my old dipper and now rely on a contractor with a mobile dipper that's much safer to use. After dipping the sheep he then properly disposes of any spent dip that is left. Until a safe alternative can be found, I intend to use such chemicals sparingly.

I keep my flock free of parasites by quarantining all newcomers to the farm for several weeks. That way I am sure they are free from disease. More recently, I have started to breed my own flock replacements so there should be no risk of introducing disease to the farm.

Livestock can be a major source of disease. Cuddly lambs and sheep can give you the horrible skin disease called orf. Sheep also catch toxoplasmosis from cats that causes them to abort and also leads to ME-type symptoms in humans. Most serious of all is a disease of ewes called enzootic abortion that can lead to women miscarrying. If a woman is pregnant, it's vital that she stays away from sheep at lambing time. She shouldn't even allow her husband or children to bring contaminated clothing into the house.

Calves can give you severe diarrhoea through bugs like deadly E. Coli, cryptosporidia and salmonella, but don't worry, we pasteurise milk to kill off all those germs. It's also inevitable that you will catch ringworm if you work with cattle. It's caused by a skin fungus and is irritatingly itchy. Leptospirosis and brucellosis are all potentially transmissible from cows. That's why all adult cattle are now routinely blood tested to identify carriers. Fortunately, the lethal killer anthrax is now rare. Elsewhere, there are risks of Weil's disease from handling hay bales contaminated with urine from infected rats.

Having survived all those diseases there are the animals themselves, as even docile beasts can cause serious damage. Most folk know to beware of a bull, but few realise that cows can be just as dangerous. Being lighter, a cow can move faster than a bull. Worse, they often give little warning and are often more persistent in their attack.

Cows are usually at their most dangerous just after they calve. It's their natural instinct to protect their newborn calves. Some paw the ground with their forefeet and give menacing warning bellows. That's the signal to stay clear. Within a few days, when her calf is stronger, she'll be as placid as can be. Others stand dolefully by as you check the newborn calf and suddenly charge without warning. I have had a few frights with freshly calved suckler cows playing that trick. If they manage to knock you down, they will gore you with their head or crush you with their brisket.

I was once very nearly killed by a cow defending her calf. She was a small, Aberdeen Angus-cross aged five that had just given birth to her calf first thing in the morning. As usual, I loaded her calf into a wheelbarrow to give it a hurl to a straw-bedded pen where it could learn to stand and suckle. That wee cow anxiously followed me to the pen sniffing her calf all the way to make sure that it wasn't being harmed.

After breakfast, I prepared the paperwork in readiness for fitting official ear tags to the calf. It's all part of the bureaucratic process of modern livestock farming. Every calf has to have a passport that records details such as its date of birth, sex, breed and official ear-tag number as well as the ear-tag numbers of its mother and father. You fill in these details on an application form that's sent to the British Cattle Movement Service. They record those details on a central computer and then issue a passport. It's illegal to move cattle without a passport so they can't be sold and become worthless. There's also a strict time limit and if you fail to register a calf within twenty-seven days of its birth you are denied a passport.

I do everything the day the calves are born. Snag was, when I went to fill in the form I realised I hadn't looked to see what sex the calf was. So I nipped up to the calving shed and, without thinking, climbed into the pen to lift the calf's leg to see if it was a bull or a heifer.

Suddenly, the cow charged me to protect her calf. She repeatedly rammed me against the wall with her head and I was trapped in a corner. Every time she charged, my head banged against the concrete wall and I really thought she was going to kill me. Then I had a stroke of luck: another cow made a noise that distracted her for a moment and I managed to escape. That left me badly bruised with a sore hip and shoulder, but I soon recovered. Sadly, the cow was slaughtered that autumn after she had reared her calf. As I often work alone, I couldn't risk being charged again the next time she calved. You just can't be too careful with freshly calved cows.

In fact, all farm animals can be dangerous and have to be handled with caution. Cattle get nervous when penned. Some kick with their hind hooves as you approach. Worst of all are young cattle, such as half-grown suckled-calves, that can kick with lightening speed. Cattle can also accidentally stand on your toes, crush you against walls or knock you over as they rush by.

Sheep are equally dangerous. Rams can charge and break your legs or cause serious head injuries if you are knocked down. Horned rams, such as the Scottish Blackface, can accidentally damage your legs and knees as they run past you in the pens. It's not uncommon to get a broken nose, broken teeth or a black eye from a struggling ewe or lamb that jumped in the air and inadvertently head-butted the shepherd.

Pigs, particularly sows defending their young, have a nasty bite, and the long tusks of a boar can inflict serious wounds. Horses bite and kick. Farm cats scratch and bite. So do rats. Excited collie dogs will bite, particularly when forcing a bunch of sheep or cattle into pens. Cockerels, stag turkeys and geese will fly at you to peck or scratch with their spurs.

A farm can be a very dangerous place. Even my wife has been known to attack with a broom after spying muddy boot marks on the kitchen floor!

It's a fact of life that we all grow older and a glance at the mirror will confirm that. Everywhere I look people seem to be younger than me, except when I go to the market. The average age of farmers is fifty-nine and rising. Previous generations often held on to the purse strings until they were well past their seventieth birthday. That was frustrating for sons who patiently waited to have the reins handed over to them.

One of the problems with agriculture in the past was that youthful enthusiasm gave way to middle-aged caution. While father may have been well past retiring age, that really didn't affect the running of the farm. Sons did all the hard graft while father went to market and tended to the business. Things have changed a lot over the years and most of us now recognise the importance of handing over to the next generation sooner rather than later.

Sadly, few youngsters now want to farm. They have seen their friends get better-paid jobs with shorter hours. They have voted with their feet and abandoned farming. So, more and more, elderly farmers are being forced to carry on working their farms.

Another problem that is forcing farmers to continue working is that few of us have adequate pension provision. We reinvested any surplus cash in the farm. With house prices booming, selling up is no longer an option for tenants and many are forced to continue to live in the farmhouse and eke out a living from the land.

So an announcement by the Health and Safety Executive (HSE) that a quarter of fatal farm accidents involved farmers over normal retirement age, with some in their eighties, came as no surprise. The HSE pointed out that even relatively minor accidents become life-threatening for the elderly. We all tend to forget that we become less agile as we get older. I now think carefully before working with cattle to make sure that it's done safely. It's a sobering fact that an injury could make it impossible for me to run the farm and I simply can't afford the wages for my replacement.

Many of my non-farming school chums have taken early retirement. I don't envy them because I enjoy farming, but I do worry about what the future holds.

20

High and lonely places

At one time it took a fair number of men to work a farm. Depending on the size and type, there may well have been a foreman, known as a grieve. On bigger farms, there was the first horseman with his lads, right down to the fifth, or even sixth, horseman. Each looked after and worked a pair of Clydesdales.

Dairymen, cattlemen and pigmen all had their own jobs and their own livestock. 'Jack of all trades' was the orraman who had to turn his hand to anything. Alone among them was the shepherd. While the rest often worked together, and maybe ate and even slept together in the bothy, the shepherd took himself off to high and lonely places. Out in all weathers, he tended his sheep with only his collie dogs for company.

Not a lot has changed; to this day, shepherds are a class apart. No man dictates their hours, for shepherds do their job in their own time.

They often have to start at daybreak to gather sheep before the heat of the sun makes the steep braes unworkable for dogs. At lambing time they might well toil from five in the morning until ten at night. It isn't uncommon for them to work well over a hundred hours a week. At other times of the year there is little to do and twenty hours a week is all that's needed to do the job properly.

About ten years ago farmers' representatives and the farm

workers unions set about simplifying conditions of employment. The old agricultural wages schedule ran to about eighty pages, listing the various rates of pay and employment conditions for the different categories of worker. With so few of us left on farms most of us do whatever is required and there's no demarcation.

The new wages schedule recognises this and has only three categories: for part-time or casuals, full-time workers and qualified craftsmen. The idea was to be flexible with rates of pay so that those who worked long hours were properly rewarded. That didn't suit those shepherds who had built up a tradition of trust with farmers. Just as farmers hadn't dictated working hours to shepherds, many couldn't be bothered to fill in timesheets. In other words, as far as a lot of shepherds were concerned, nothing changed.

I remember one old shepherd who had herded the same hills all his life and had been born in the cottage that he lived in till he retired. His father before him was shepherd on the same farm. As soon as my friend was strong enough to keep up with his father he followed him round the hill. He regularly skipped school to help with gatherings, or to attend sales and local shows. When he left school, his father taught him his craft and he took over responsibility for the flock when the old man retired. The hills that he worked on had belonged to the same family for four generations.

That's one of the great differences between the farming workforce and industrial workers. Shepherds, tractormen and cattlemen often build up long-term relationships with farmers because of the nature of the job.

Those with farms large enough to justify staff only hire a few at most and that means working closely with the boss and getting to know him. Unlike factories, there's seldom any demarcation on a farm. When it's the boss's turn to feed the livestock at weekends, he will do the same chores as his stockman. As a result, strikes are non-existent, as each has a better understanding of the other's point of view.

That's not to say there aren't rows. Farm workers often fall out with their boss and move onto other farms but, by and large, most good men settle into the ways of their employer. As a result, farm workers regularly receive long-service medals. Of course, there are some bad employers about, but fortunately they are few and far between. I know of one who can't keep shepherds longer than six months to a year. As well as being an objectionable man he doesn't know much about sheep, yet tries to dictate to his shepherds. The shepherding community is a small one and all the good shepherds know to avoid that farm. In a small farming community, that's maybe another reason why most farmers treat their workers fairly and with respect.

What we Scottish farmers refer to as the 'term days' fall at the twenty-eighth of May and November following Whit and Martinmas and are traditional dates to start or end farm tenancies and staff contracts. At one time the country roads were busy with farm workers flitting on the 'term'. Some would have changed jobs for better pay or conditions, while others merely moved on for the sake of it. 'It's time for a change,' was a common reason.

Years ago those flittings were done by loading a livestock lorry, or a tractor and trailer, with the household furniture and goods. Despite being carefully stowed, roped down and covered with a tarpaulin to keep the rain off, it was amazing how much got damaged. Rattling up a bumpy farm road broke china and chipped furniture, and it wasn't uncommon for the odd chair to fall off on to the road. Nowadays the job is done professionally with furniture-removal vans.

Then there was the scrubbing down of the new farmhouse or cottage and all the upheaval of settling into a new home. The May term was also the traditional date for shepherds planting out their gardens. Most had been busy lambing and anyway, high-lying land, where most shepherds live, doesn't warm up until the end of May. It's amazing just how well a garden can grow despite being planted so late.

Fortunately I have only done a farm flitting once and it was hectic. While the women organised the house, the men had the job of moving lorry-loads of sheep, cattle and farm equipment as well as all the furniture. Never again, I said at the time, forgetting that some day I will have to retire and move out.

'You have to pay for your education' was one of my father's favourite sayings. What he meant was that learning from your mistakes can be expensive. There's no doubt that education and sound training are the keys to a successful career in farming, as they are in other walks of life.

Snag is, it costs so much to give a lad an apprenticeship nowadays, and who has the time or patience? When I was a lad there were always lots of men working on the farm. With so many about, someone always seemed to have the time to show a lad the right way to go about a job.

Nobody seemed to be in such a rush in those days. Not that the jobs were sophisticated, unlike many of today's more technical tasks, but some simple jobs still required a lot of skill. Making hay, building a cartload of bales and stacking them in the shed were all skilled jobs that we gradually learned through repetition. Then, before we knew it, we were in our twenties and competent tractor drivers, stockmen or whatever. Mind you our apprenticeship started at a very early age. I worked on the farm every school holiday I can remember and most farm lads at the time were of some use to their father by the time they were nine or ten.

On a modern farm there is so much sophisticated equipment to maintain, so many drugs and chemicals to use and so many high-tech systems. With fewer helping hands, everyone is rushing about with little time to explain it all to youngsters. That's why today's farm lads have to be professionally trained.

As I have said, many farmers now run their farms single-handedly without any hired workers at all. That's not so difficult with modern equipment and farming methods, but the problem is arranging time off. In the old days, when there were plenty of workers about, the farmer was hardly missed. Days away, weekends off, or even holidays, were not a problem and the workers enjoyed a break from their boss!

Even those who didn't have staff could rely on finding a local farm worker on holiday or retired farmer to help, but, with fewer of them about, that's all changed. Arable farmers aren't so badly affected, as their work tends to be seasonal, allowing time off between, say, sowing and harvest, but those with livestock find it much harder to get away. Animals have to be checked morning and night, and while beef cattle or sheep may not need to be fed during the summer, pigs and poultry certainly do.

Dairy farmers are worst affected, as their cows must be milked at least twice a day. Even trying to arrange a half day to go to a wedding can prove to be a nightmare. There are agencies that supply relief milkers; sometimes self-employed specialists, but mostly farmers' sons looking to earn extra cash, and they don't come cheap.

Still, all work and no play is no fun and, apart from the farmer needing a break, there are also his wife and children to consider. After all, they are just as entitled to a weekend away with their father as any other family.

21

My favourite cow

A farmer's job is to look after his livestock, not look for them. Time spent looking for strays is wasted and unproductive. Sheep are expert escape artists, the Houdinis of the animal world. They can jump over a five-feet-high drystane dyke with the same ease as a deer. Once one of them learns that trick, the rest aren't long in following its example.

Unfortunately, not all sheep can jump a dyke without dislodging the top stones, so a flock of jumping sheep can soon destroy the dykes on a farm. That is why jumping is a cardinal sin on my farm. On the rare occasions a sheep does jump, I make a point of catching and selling it.

Sheep are also experts at crawling under gates or squeezing through gaps in hedges or small holes in fences. It can be a constant battle, particularly in the hungry months of February and March, to keep them where they belong.

Cattle on the other hand don't have the same wanderlust. Bulls sometimes break out to fight with other bulls, or for a session of passion, but most cattle stay contentedly at home in their fields.

Biggest cause of cattle straying on our farm is hillwalkers not shutting gates properly. So it came as a surprise when one of my cows learned to stray onto a neighbouring farm.

All my cattle are counted and checked every morning when I simply give each lot a shout to gather them up. Twice in one week a cow was missing from one of the lots. After fruitlessly searching over undulating hill ground for a sick cow that never materialised, I was relieved to find they were all present when I recounted them. Both times I assumed I had somehow miscounted until my neighbour telephoned to say that he had seen one of my cows crawling under a water gate and returning unseen as I looked for her.

Water gates are designed to prevent livestock straying when they wade up a burn. They are simply wooden gates suspended from a telegraph pole or wire rope so that they swing to allow the flood-waters to flow under them. That crafty cow had worked out that all she needed to do was to put her head down and keep walking, and the water gates would give way in front of her. Sadly, I had to sell her before she taught the rest of the herd the trick.

Then there was the time a heifer appeared to vanish into thin air. Heifers are young females and aren't called cows until they're about to give birth to their second calf. That heifer was fifteen months old and was running with fourteen others of a similar age when they were badly spooked one Sunday morning.

I found eleven of them in a silage field, while another three were in an adjacent field. They had obviously stampeded during the night and most of them had jumped over a gate in the corner of their field, while three had hurdled a fence. That metal gate was badly mangled as one of the heifers had got her back legs caught between the spars as she jumped over. As she struggled to free herself the other beasts had panicked and jumped over her, buckling and twisting the gate. That poor heifer was left with a very sore back-leg and had a bad limp for the rest of the summer.

Rounding up the escaped cattle wasn't an easy task as they stampeded every time my daughter and I approached them. Eventually, we managed to get them in a secure field and left them to settle down while I went to look for the missing one.

I searched everywhere to no avail!

Cattle are sociable creatures and don't like being alone. Stray beasts soon spy other cattle and join them for company. I reckoned if the heifer had been hiding in the woods surrounding my farm she would soon have tired of being alone and joined one of the various groups of cattle grazing nearby. Sadly that didn't happen and none of my neighbours spied her either.

The police were notified in case she turned up on a busy road or railway. I also alerted the local gamekeepers because they patrol all the hidden nooks and crannies and they were probably my best hope. Despite everyone being on the lookout, nothing was reported.

It's illegal to move or sell cattle without a passport that corresponds with their official ear-tag numbers. That made the heifer worthless because her passport was securely locked away in my office. While she couldn't be sold legitimately, she could be shot and butchered. The simplest explanation was that someone had shot her during the night and spooked the rest. I feared the worst and suspected that she had ended up in a freezer. Such criminal activity is rare but seemed the likeliest explanation for a beast vanishing into thin air.

Fortunately, I found her one morning later that week amongst a bunch of cows and calves and I will never find out where she had been. Only problem was that like the rest of the heifers that had been spooked, she had become very timid. Whenever I approached those heifers they pricked their ears in the air and moved away to a safe distance. I knew from experience that if I were to make a sudden movement they would have stampeded.

Timid or wild cattle are no use to anyone. It's important that cattle are calm and placid by nature, because we have to work with them. Not only can wild cattle cause a lot of extra hassle by stampeding or straying, but they also damage gates, dykes and fences. Nobody likes having to work with wild cattle either when they are calving or when carrying out routine tasks like vaccinations or reading ear tags.

To calm those heifers down I introduced a couple of old cows to the bunch. Nothing frightened those quiet old girls and, as they placidly stood their ground, it reassured the rest. Better still, I started feeding them delicious cattle cake in troughs every morning. Greedy cattle can't resist that and gladly run towards you for the feed. As they munched away, I quietly talked to them to reassure them. They soon quietened down again and started coming to me to lick my coat or have their noses rubbed. We had become the best of friends again.

Good friends are hard to come by, but I saw a fine example of friendship about five years ago. Every year I cull about ten cows from my herd of seventy. Some are sold because they are getting too old, while others are infertile. The main reason for cows being culled is a disease called mastitis. It affects the udder and dramatically reduces milk yield so that the cow struggles to rear her calf properly. Mastitis is spread by flies, which are common on my farm, as it's surrounded by woods.

Infertile cows are sold in the early summer as soon as they are fat. The rest are culled in the autumn once their calves are old enough to be weaned. That year, there were four cows with mastitis to be culled. As usual I ran them separately with their calves in a small field. That group of cows and calves spent the summer with a couple of yearling bullocks that weren't suitable for sale in the spring. One of them made a snoring sound all day long while the other had a misshapen head that had earned him the nickname of 'Elephant Man'. The previous winter that bullock had inquisitively stuck his head through a gate to sniff the bull in the next pen. That provoked the bull into crushing the calf's head, which caused a permanent swelling on one side.

Both those misfits were to be fattened the following winter, with the best one ending up in my deep freezer. Elephant Man became close friends with a little heifer calf and they spent most of their time together. I often saw that big ugly bullock lick the little calf as if he were her mother.

One Monday evening I rounded up the four cows and their calves and put them into a shed leaving the two bullocks behind in the field. The idea was to wean the calves and send the cows to market.

As usual, the calves spent the night roaring for their absent mothers and next morning I found Elephant Man standing outside the shed. He had jumped four dykes to be with his wee pal. He was a quiet bullock, so I soon walked him back to his field, but ten minutes later he was back outside the shed roaring to be with his mate. The only solution was to let him into the shed and put him in a pen beside his pal so that he could keep an eye on her.

Just as Elephant Man came to grief by sticking his head through a gate, one of my bulls got his head stuck after poking his nose where it didn't belong. My cattle are fed through a feed barrier. It's a sort of fence, with railings of tubular steel spaced fourteen inches apart. That lets cows stick their heads through to feed, but prevents them escaping from the pen. Bulls' heads are too big, so they feed through a self-locking yoke, which not only is wider but also can catch and restrain them. I simply turn a lever so that when the bull puts his head through it forces a bar firmly against his neck and securely holds him in place.

You should never trust a bull. When I have to work in the pens where there are bulls – for example when I'm bedding them with straw – I always have them safely held by their self-locking yokes. The bull in question had spent four winters in his pen and had never attempted to put his head through the narrow gaps of the feed barrier. He always used the yoke to feed through.

You've guessed it! For some reason he decided to stick his head through one of the fourteen-inch gaps in the main part of the feed barrier and got well and truly stuck. At first, I thought it would simply be a matter of clapping my hands and shouting to encourage him to pull back a bit harder. When that failed, I tried twisting his head at an angle to see if that would do the trick.

Have you ever tried twisting the head of a bull weighing well over a ton? It was ineffective to say the least. My wife then had the bright idea of covering his head with soapy water to see if that might help him slip his head back. After wasting a lot of washing-up liquid we concluded that wasn't going to work either.

Finally, I decided to cut a tubular steel bar at the bottom of the feed barrier to make the gap bigger. First, I tied his head securely with a rope halter before cutting the steel with a large angle-grinder. That's when all hell broke loose! The high-pitched whine of the grinder and the shower of hot sparks terrified the bull. As soon as I had cut through the steel I removed the halter and the bull bent the steel for me as he pulled his head back. Free at last!

Livestock aren't always hassle and can sometimes pleasantly surprise you, like the time a two-year-old Aberdeen Angus heifer gave birth to her first calf. I found her at six o'clock one morning fussing over the new calf, which was a wee heifer and slightly premature. As it hadn't learned to stand my first task was to lift it into a wheelbarrow to give it a hurl to another shed where I had prepared comfortable straw-bedded pens. That little heifer calf was soon onto its feet and suckling its mum. When I checked them at bedtime they were contentedly lying together on the straw.

Imagine my surprise the following morning when I discovered that she had given birth to another calf through the night. Her second calf was a bull that must have been born at least seventeen hours after his sister. Normally I expect about 3 per cent of my cows to have twins but, after forty years of working with cattle, I have never had a heifer with twins.

The only snag with my bonus was that the heifer calf was probably a freemartin. That's where the male hormones from the bull calf interfere with the development of his sister in the womb so that she becomes infertile. Despite that, she grew to be a fine beef animal like her brother and fetched a good price at market the following spring.

That year's good fortune contrasted with the poor calving of 2005 when I lost more calves than usual. It wasn't that there was a problem with disease, but rather that I was unlucky. Three calves suffocated at birth as a result of the birth membrane covering their nose and mouth. That happens from time to time, but thrice in a year is unheard of.

Another snag was the higher proportion of bigger calves, with the result that I had to assist quite a few. On three occasions the vet was called in case we had to do a Caesarean section. Those big calves were the result of better-than-usual silage, which meant that my well-fed cows nourished their unborn calves better. Typical of that year was the following incident when there was only a handful left to calve. One of them, a two-year-old heifer calving for the first time, obviously needed assistance. I could see the unborn calf's protruding fore feet and they were larger than usual. That usually means a big one is on the way.

My wife and I walked her to the farm buildings where I tied her by a rope halter. Then I fixed two short ropes to those two front legs and fixed those ropes to a calving jack. It's a ratchet-type device that firmly pulls the calf out. Slowly but surely we eased that big calf out of the heifer alive. It was a bull calf and we left its mother licking and fussing over it. Later on, I found that it hadn't learned to stand. Obviously a calf that can't stand can't suckle either. So I hand-milked the heifer and fed the milk to the calf using a stomach tube. It's passed down the calf's throat so that milk can be poured directly into the stomach to avoid any risk of choking.

After several days of feeding the calf by stomach tube three times a day, my wife and I eventually trained it to stand. The snag was, it always had to be helped to its feet and it never learned to walk or suckle its mother. On the ninth day we reluctantly had to accept that the calf was abnormal and I got the vet to put it to sleep. My wife was very upset, particularly after all her tender care. That incident reminds us that, as with humans, there are those who live only briefly and then there are others who are blessed with long lives.

I will never forget a local joiner working on a nearby farm who made that point very eloquently. He had just finished a job and everyone was chatting as he loaded up the van to go home.

He pulled out a measuring tape and declared that man's allotted time was three-score years and ten. So he drew out seventy inches of the measuring tape and asked his apprentice how old he was.

The apprentice was seventeen so he wound the tape to fifty-three inches and assured him there were a lot of inches left. Another in the squad was forty-five so he drew the tape back to twenty-five inches and pointed out that more than half were gone. The joiner then solemnly declared that he was sixty-four, making the remaining six inches look very short indeed. On that sobering note he reminded the group to live life to the full!

Farm animals, like people, are all individuals. There are fat ones, thin ones, wee ones and big ones. Some have white heads while others may be black or speckled. They can be hairy or sleek, have big ears or wee ears, and on it goes. When you work with animals every day, you soon get to know them as well as your family or friends, and that's when you begin to notice their different temperaments.

Some are timid while others are bullies. There are greedy animals, inquisitive ones, intelligent ones and some that are plain daft and always getting in a predicament, like becoming stuck in a ditch or tangled in briars.

For some years my favourite cow was 115, so it was a sad day when the vet had to put her down. Number 115 was a friendly beast that loved to have her forehead scratched. Whenever I went into her field she came running up for a good scratch. In winter she would poke her head through the feed barrier as I walked past. If I was in a hurry and ignored her she would bellow disapproval until I made amends.

That all stopped one January a few years ago when she began to ignore me. For some time, I had noticed that her behaviour was becoming more and more eccentric: she started standing in the same part of the pen at feeding time; then she took to walking slowly round the outside of her pen. Eventually, I became concerned when I noticed her salivating slightly. On closer examination I got the impression she had become blind in one eye. So I decided to take her out of her pen and put her in the sick bay.

When my wife and I started to move her, 115 went crazy and charged at me. Once out of her shed she had a fit and fell onto her side with her legs thrashing wildly. After an hour she had calmed down and recovered enough to get back onto her feet.

Sadly, the vet diagnosed a brain tumour and decided that the kindest thing was to have her put down. I restrained her with a halter and reassured her as the vet went about his deadly business. It was a quick and painless end, but I still have fond memories of my old friend.

Then there was Old Casanova, a seven-year-old Suffolk tup that got couped on his back. He was unlucky and couped in the early evening just after I had done my last rounds. So he had lain upside down all night and died in the early morning just before I got to him. He is sadly missed. Although he wasn't much to look at, he was, as his name suggests, very good at his job. He loved all ewes and most ewes adored him.

In the autumn I expect most tups to serve about thirty-five to forty ewes. A bunch of 150 ewes will normally run with four tups. When I check the sheep in the morning each tup will usually be working on a couple of ewes. Except for Old Casanova, who always had at least half a dozen in tow waiting for his amorous advances.

Woe-betide any tup that decided to steal from his harem. Whenever another tup approached he would put his head down and threaten to charge. That was often enough to scare away the rival, as most tups knew from experience that a head-butt from Old Casanova was a painful affair.

The autumn before he died he served more than his share of ewes and then went on to serve another fifty older ones. In other words, he probably fathered at least 150 lambs the following spring. I reckon that he must have fathered at least seven hundred lambs in the six years he was around. I bet there were a few ewes with tears in their eyes when they heard that the old boy had passed away!

Pet lambs can be real nuisances. Throughout the lambing they are usually being fed milk from a bottle. They are lambs that have been found abandoned or may have been lifted and brought into the sheds because they were either twins or triplets and their mother hadn't enough milk to rear them. Occasionally, their mother has died.

As the lambing progresses, we foster those pet lambs onto ewes that have lost their own, or onto those that have only one lamb but with enough milk to rear two. By the end of most lambings the pet lambs are gone, but occasionally there are one or two that didn't get foster mothers. In such cases we continue to bottle-feed them with milk until they are eating reasonable quantities of special concentrated lamb pellets. At that stage we wean them, usually at four to six weeks old.

Pet lambs naturally assume that the person who bottle-feeds them is their mother and run after them as if they were ewes. Having no fear of humans they get up to all kinds of tricks. Being tiny they quickly learn to wriggle through, or under, gates, fences and hedges, often in a bid to get near their 'mother' – my wife!

Soon they have mastered the art of escaping from anywhere. Garden flowers and vegetables are a favourite nibble to steal and that drives my wife mad.

A friend once had a pet lamb that loved to come into the kitchen and sit with his children as they watched television. As it got older it learned to sit in my friend's armchair to watch the telly. That made a mess of the chair so it was banished to the fields and the farmhouse door kept firmly shut.

My neighbour had a ewe that got into a real predicament as a result of a notorious escape. He has a disused quarry next to his land that is fenced-off to prevent sheep and cattle falling off its sheer, rock face. A group of his sheep spied a hole in the fence and wriggled through it into the quarry. They found the rewards well worth the effort. Having never been grazed, it was covered with juicy bits. One adventurous ewe grazed along a rocky ledge some forty feet above the rest and was soon joined by a couple of her pals. Snag was, getting down onto that ledge was easier than climbing back up it.

After a couple of days, two of them did manage to scramble back to safety, leaving a terrified companion behind. You would be amazed how many folk noticed that stranded sheep as they drove up and down that road. My neighbour had many anxious phone calls alerting him to the plight of the silly animal. He thought about lowering himself down the rock face on a rope to rescue it, but decided that was too dangerous and not worth the risk. He even thought about seeking the help of professional climbers or hiring a rescue helicopter with a winch.

Fortunately the sheep solved his problem by rescuing herself. It may have been hunger that prompted her, or maybe she got tired of being on her own and wanted to rejoin her mates. Whatever her motivation, she plucked up enough courage to climb back to safety. My neighbour found her safe and well with the rest of the flock. He has now repaired the fence surrounding the quarry so there shouldn't be a repeat of that incident.

I just hope the ewe learned that, while grass on the other side of the fence may be greener, it might not be as safe.

22

Changing times

If I could talk to my animals, I am sure they would tell me a lot about the changes they have seen over the years. The way tractors have taken over from the mighty Clydesdales for example. Every farm still has traces of them: from rusting horse-drawn implements lying in corners of the yard to pieces of harness hanging from rafters, to those massive horseshoes that are regularly turned up by the plough. Nowadays the majestic beasts that remain are found only on farms owned by enthusiasts. They turn out their magnificent charges beautifully for agricultural shows or ploughing matches.

Other animals of mine would reminisce how most farms kept at least one pig, if not a few sows. Gradually those pigs – fattened and killed on the farm for home use – disappeared, and pigs fell into the hands of specialists. Large intensive units were developed that are capable of keeping several hundred sows. Recent unfair competition from imports has destroyed our pig industry and there are now a lot less pigs being kept. It's almost inevitable that many more Scottish pig farmers will go out of business. Scottish pigs could become as rare as Clydesdales, and treats like Ayrshire bacon become a thing of the past.

Hens, chickens and turkeys would also tell me a tale of intensification that has led to their virtual extinction in the farmyard. Poultry are now vertically integrated: breeding, fattening, feed milling and slaughtering are all carried out by one large company. Such companies are more efficient and have the power to negotiate better terms with supermarkets.

At one time many mixed farms milked cows, as the monthly milk cheque guaranteed a regular income. It's not so long ago that butter was churned in the farmhouse and delicious cheeses were made for local markets, but strict hygiene regulations have made traditional farmhouse cheese very hard to find.

Dairying is also becoming a specialist business where large herds are efficiently milked in computerised, stainless-steel parlours. I suppose my beef cows would have the most to say to me. They must be amazed that we not only shoot unwanted bull calves out of dairy cows and older, lean ewes but also incinerate older cattle rather than send them to be butchered. I can hear those cows chortling at all the troubles we have brought on ourselves by these unwelcome changes.

One of the biggest changes has been the increase in rural crime. I never gave a thought to security when I moved into my farmhouse thirty years ago as a recently married young man. The farmhouse doors were never locked, cars and tractors always had the keys left in their ignition and there wasn't a padlock on the farm. How things have changed.

Nowadays farm vehicles are securely locked up, with anti-theft devices activated, and then the garage doors are securely padlocked. When I go out to check my cattle at night I set off so many security lights that the farm steading looks like Blackpool illuminations. My movements are recorded for posterity on closed-circuit television cameras and, like a jailor, I am forever fiddling with a bunch of keys to find the one that fits the lock I am about to open.

Sadly, remote farm steadings have become rich pickings for thieves and we all have to be security conscious. Farm workshops are full of valuable tools and there is always an all-terrain vehicle

(ATV) to be found somewhere. There must be a massive market for dodgy ATVs judging by the number that are stolen, although I have never heard of the police recovering one. They seem to vanish into thin air once they are stolen. Not only do thieves add insult to injury by stealing the security devices that protect the machine from theft, but also they return a few weeks later to steal the new one replaced courtesy of an insurance policy! Aluminium livestock trailers are another favourite and I had one stolen a few years ago.

In my opinion, the worst crime of all is when livestock is targeted. That really annoys me because, invariably, it is farmers stealing from fellow farmers. Sadly, in my opinion, modern police forces are of little use in preventing rural crime and it's up to farmers and other country dwellers to help themselves. Everything of value has to be securely locked up, security lights fitted and closed-circuit television cameras installed at strategic points.

Electronic-tracking devices can be fitted to more expensive equipment whilst simpler ideas – like painting registration numbers on the roof of livestock trailers – are cheaper and just as effective. I know of a farmer nearby who has one of the best security devices. It's a collie bitch that lies quietly out of sight until she sees a stranger moving about the farm steading who hasn't gone to the farmhouse first. With considerable stealth, she quietly stalks the intruder before taking a firm grip.

After years of nipping at beef cattle she has learned to be fast and accurate, but she also knows that a fleshy, human backside is safe to hang on to, unlike the kicking heels of a beast. It's amazing how word of that wee collie bitch has spread throughout the community! Mind you, there are more than collies keeping a watchful eye on things.

Townsfolk can be forgiven for thinking that living in a farmhouse without next-door neighbours is a lonely way of life. To those used to being surrounded by flats, houses and busy streets, a remote house in the country must seem very isolated. Truth to tell, it often seems to me there are more isolated and lonely folk living in towns or cities. It's easy to get lost in a crowd.

Despite our relative isolation, country folk tend to form tight-knit communities that don't miss much. Untrained eyes might not realise that others are watching. Farmers, shepherds and gamekeepers often go about their daily work unnoticed by townsfolk, and we see a lot more than you think. Many pieces of gossip are started by a chance observation. There again, the offer of a helping hand or a visit to cheer someone up may be prompted by noticing something out of the ordinary. Mind you, there are more than human eyes watching over you in the countryside. Few people realize that farmers are being spied on by a government agency using satellites.

European Union rules insist that member states must inspect a minimum of 5 per cent of the land on which subsidies can be claimed under the Common Agricultural Policy. Since 1992, our political masters have been using satellites to check that farmers abide by the rules. Images from the high-resolution cameras on board the satellites are so good they can measure distances as small as a metre. Everything from the area under crops to the width of margins round the edge of fields, footpaths and hedgerows can be measured accurately. Those photos are used to check that we comply with the rules that go with our claims under the various support mechanisms and environmental schemes. If officials spot a discrepancy, they follow it up with a farm visit.

Everything in the countryside is under observation. Closed-circuit television cameras in towns are easily spotted, but satellites are invisible. Next time you are in the countryside, remember that a spy in the sky is watching.

23

I hope I never see the like again

During the outbreak of foot-and-mouth in 2001 I kept a diary for *The Scotsman*.

Tuesday, 20 February

Took delivery of the best load of feeding wheat straw I have ever seen. Mind you, at £64 per ton it would need to be. Having said that, I would rather pay a bit more for the best than deal with cheaper, mouldy stuff.

Turns out the load came from a Northumbrian farm so I had a long blether with the lorry driver about the suspected case of foot-and-mouth down south. Agreed that it will probably turn out to be a false alarm.

Thank goodness it's well away from here.

Wednesday, 21 February

I was away to Castle Douglas for a one-day training course to learn how to value standing hardwoods. It's all part of being appointed by the Forestry Commission to its regional advisory committee. It was a pity about the early-morning start. I bet nobody else on the course had to feed 70 suckler cows and their followers as well as 650 sheep before they arrived.

Listening to Radio 2 in the car I learned that foot-and-mouth has been confirmed. Fortunately MAFF vets seem to be on top of the job and will easily control it with all the modern tracing systems we have in place. I'm disappointed that the virus has got into the country. It was probably some dozy tourist illegally bringing back meat products in their luggage that has found its way into a pig.

Thursday, 22 February

I had another early rise so that I could give the cattle extra silage to allow me to attend a meeting in Glasgow of the E. coli 0157 task force. Fellow committee member Professor Hugh Pennington was constantly on his mobile giving statements to the press about foot-and-mouth. That man definitely thrives on publicity.

There was a lot of talk among the vets on the committee about the suspected case in Fyvie; they were speculating that it might just be symptoms associated with the recent cold snap. You would have thought that highly trained vets could tell the difference between foot-and-mouth and frostbite. According to the vets the virus has lots of ways of being spread, such as on the wind or even on a load of straw.

Friday, 23 February

I am getting a bit nervous about that load of straw, so I telephoned the merchant to find out where it actually came from.

It seems it came off a farm that also keeps pigs. The merchant realised that I was having one of my panic attacks and reassured me that it was at least fifty miles away from the outbreak. Fifty miles is nothing to this virus. The loaded wagon probably passed within a few miles of the outbreak and picked up every windborne virus in the area.

I am toying with the idea of selling that straw cheap to someone I don't like.

Saturday, 24 February

According to several radio and television reports, and the newspapers, the virus has got into the network of livestock dealers. The potential for the disease being spread far and wide is now very real.

I have never really cared for livestock dealers and regard them as parasites that do little for the industry. There is an obvious need for procurement agents to source animals for the various abattoirs scattered around the country. I can even see the need for agents to arrange for farm-to-farm delivery of breeding or store stock. After all, not every farmer has the time or inclination to regularly attend markets, but I can see little point in dealers who hang about markets to pick up livestock that are then tried speculatively through another auction mart. Apart from the unnecessary stress to the animals involved, the process does nothing for the industry.

The argument that livestock dealers are another important outlet falls flat when you recall that their presence at markets still didn't prevent old ewes collapsing in price to the point where you couldn't give them away. They are the 'Del boys' of the farming industry, who tend to deal in the poorer quality animals most likely to be carrying disease. It's about time they were outlawed.

Sunday, 25 February

Would you believe it? I saw a man walk across my hill this morning with two dogs that weren't on leads. For once in my life I bit my lip and avoided confrontation. It's incredible to think anyone could be so thoughtless after all the recent publicity about staying clear of farmland.

Monday, 26 February

Now up to twelve confirmed cases. Seems that the Devon dealer traded out of Longtown mart, which is near Gretna. That could prove to be fatal as there is an unwritten law that every old ewe in Britain must pass through Longtown at least once.

Longtown is getting dangerously close to home and there are a lot of local farmers who sell there. I also heard that a dealer in Lockerbie has a suspect and that there are a couple of hundred farms under supervision.

Disinfectant is now in short supply.

Tuesday, 27 February

There was a storm through the night and blizzards. A chimney pot on the farmhouse blew down and brought a dozen slates with it. There were various roof sheets, Yorkshire boarding, gutters and overhead cables blown down.

I stayed in the house till after lunch as such debris blowing about would have made feeding stock a dangerous task. I had muesli for breakfast thanks to power failure. When the power was restored I learned from television reports that we are now up to sixteen confirmed cases.

Fortunately I don't need to worry about disinfectant for my road end as we are now snowed in with six-foot snowdrifts. It's grand not seeing the post van laden with bills and final reminders. This northerly wind should be blowing the virus back down into England where it belongs.

Wednesday, 28 February

I had arranged to send eighteen store heifers to United Auction's Stirling mart, but that's now postponed indefinitely thanks to the movement restrictions. Thanks to the snow I couldn't have got a lorry up my farm road anyway.

Not being able to sell stock is going to make cash flow non-existent. I will need to have a 'greeting' session on the phone with my various creditors and grovel for sympathy.

I finally got to the hill and freed a dozen daft ewes that had got stuck in a snowdrift. If I get foot-and-mouth this week I will have a devil of a job gathering the hill to shoot them and the vets are going to find it even harder to get the coal and sleepers up my road to burn them.

Thursday, 1 March

Outbreaks confirmed in Northern Ireland, Lockerbie and Canonbie. It's now almost inevitable that we will see a lot more cases in Scotland before we get the outbreak under control. I saw a report on the six o'clock news that police caught a Welshman illegally moving sheep. It's the daft minority that makes these situations so hard to control.

Friday, 2 March

To hell with it all! I decided last night that if I do get wiped out with foot-and-mouth I will give up farming altogether. I don't need the hassle anymore.

I have put up with various food scares, including BSE, and struggled with low profitability, excessive bureaucracy, long hours and a hard way of life. It will take me at least five years to build up the livestock to the standard I set myself and I don't see the point at my time of life.

Saturday, 3 March

The effects of the ban on the movement of livestock start to bite and there are television reports of meat shortages. Local butchers experiencing panic buying and Smithfield traders warn that meat prices look set to rise dramatically.

Sunday, 4 March

Livestock starts to move directly to abattoirs under licence, but many farmers are disappointed by the prices being offered. Seems that abattoirs have added costs for veterinary inspections, disinfectant and slower throughput. The collapse in prices for hides and sheepskins is also cited.

Tuesday, 6 March

Feed becoming scarce due to farmers ordering more than usual to minimise the number of lorries coming onto their farms. Couldn't get my usual feedstuff so I took delivery of a couple of tons of a dearer one to tide me through.

I hear that some farmers are complaining about being overcharged for disinfectant.

Wednesday, 7 March

My ewes weren't too keen on that expensive feed so I have cut their daily ration till they adjust. I hope I can get a full load of their usual feed, as I don't fancy an outbreak of twin fever as a result of irregular feeding.

Prices for prime stock well down. It seems that the supermarkets were reluctantly forced to import beef and pork to maintain supplies and there's now little demand for home-produced meat as the market is now oversupplied.

Thursday, 8 March

A lot of complaints about carcasses left lying for far too long before they are incinerated. MAFF is now proposing to transport slaughtered animals to a rendering plant in Cheshire. Not unexpectedly, many Cheshire farmers are alarmed at the prospect of infection from such traffic.

Why do the authorities insist on building these massive funeral pyres? What's wrong with the old tried-and-tested system of burying them in quicklime? It would take a fraction of the time and avoid the ghastly situation where farmhouses have to suffer the stench of a pall of smoke for days on end.

Sunday, 11 March

Nick Brown announces that the epidemic is under control.

Monday, 12 March

Prices for pigs, lambs and cattle are still very disappointing.

Thanks to the lack of exports there are now about half a million unwanted hoggets (one-year-old sheep) overhanging the market and the Scottish National Farmers Union (SNFU) is calling for an intervention buying programme that would store them frozen.

Meat and Livestock Commission figures reveal that the United Kingdom-produced mince beef was selling for 20p per kilo more and lamb chops were up by 83p per kilo. I think I now understand what Tony Blair meant when he said that supermarkets had farmers in a stranglehold.

Tuesday, 13 March

I spoke to a friend who has had his livestock slaughtered. It was a very emotional phone call and he's absolutely devastated. He's very concerned that he hasn't the manpower to heft fifteen hundred replacements back onto his unfenced hill and reckons that even if that was possible it will take him fifteen years to get his flock up to scratch again.

Thursday, 15 March

Nick Brown announces an extended cull. It seems only yesterday that he reckoned everything was under control.

SNFU President Jim Walker breaks down with emotion at a press conference, but regains his composure and rightly backs the scheme.

Friday, 16 March

Many are angry at the prospect of slaughtering healthy animals and the Cumbrian NFU back local farmers who declare they won't allow MAFF officials onto their land.

I can sympathise with them, but sadly the sacrifice will have to be made to protect the vast majority of our livestock industry.

Saturday, 17 March

There is a fair amount of opposition to the extended cull, particularly in Cumbria. I was disappointed to see farmers on the news, congregating to publicly protest at the cull of healthy animals. With such a cavalier disregard for the way the virus spreads their animals won't remain healthy for much longer.

Monday, 19 March

As the disease is now spreading rapidly in Dumfries and Galloway it's becoming almost inevitable that my farm will either contract the virus or get caught up in an extended cull. I appointed an auctioneer to do the valuations on my behalf should the worst come to the worst.

I am also making plans for my fifteen-year-old daughter to stay in lodgings near her academy, as we don't want her prevented from sitting her Standard Grade exams through quarantine restrictions.

Tuesday, 20 March

The extended cull is postponed due to logistical problems and some of the new cases are now appearing on farms that were to have apparently healthy animals slaughtered. That reinforces the argument for such a firebreak cull and poses the question whether we should be more ruthless and extend the firebreaks even further to get ahead of the disease.

At long last the army has started to appear.

Wednesday, 21 March

It seems that a high proportion of the farms affected in the initial stages of the outbreak were not farm-assured, strengthening the argument for compulsory membership of such schemes.

Thursday, 22 March

Experts are predicting that we could see four thousand cases and this could drag on through the summer.

Tony Blair now concedes that the epidemic is going to need a lot more resources to contain it. I detect a sea change in his attitude to the farming industry and now believe that the government will enact radical changes to the way the livestock industry operates in the future. I hope that it's not just political posturing with an election in the offing and that some good will come out of this catastrophe.

Friday, 23 March

Quite a few phone calls from worried farmers. One friend in Wigtownshire has twigged that the epidemic in Cumbria is only eight miles away from him across the Solway.

Another in Argyll is becoming alarmed at the number of birds like peewits, curlews and geese migrating from the Solway coast and landing on his farm. He reckons that if tyres can spread the disease so can the big feet of a goose that has paddled amongst sheep droppings on infected farms.

The disease has crossed the psychologically important barrier of the river Nith confirming my worst fears that it will be a miracle if I am not caught up in this mess. Of more concern to me is the potential for it to break out in the large hill farms between here and Moffat where it could spread rapidly.

I am now resigned to the worst although my wife is struggling to cope with the pressure.

Saturday, 31 March

The first couple of ewes lambed.

The prospects for lamb prices this autumn look dismal with export markets closed. I may have to shoot a lot of old ewes that are normally exported to French Muslims. Perhaps it's time to develop a national scheme to collect, humanely destroy and properly dispose of them.

Sunday, 1 April (All fools day)

I wonder if landlords are reducing the rents for farms wiped out by foot-and-mouth. Those tenants who escape may also be looking for rent reductions.

Monday, 2 April

Prime Minister Tony Blair announces that local elections are postponed till the seventh of June. There is also a suggestion that vaccination is now lower on the agenda.

I am regaining my faith in the man.

Tuesday, 3 April

Oor Henry McLeish describes Scotland's foot-and-mouth crisis as 'a little problem' when speaking in the Big Apple to promote Scotland's tourism industry.

Help ma boab! Remind me to evacuate when Henry has a big problem.

Wednesday, 4 April

The number of confirmed cases tops one thousand. More worrying is the news that the welfare backlog threatens to overcome the cull.

MAFF introduced a scheme to cull animals suffering as a result of the crisis, such as ewes that were unable to be brought inside and had to lamb in muddy, exposed fields.

Nearly four thousand farmers have now reluctantly registered about 1.5 million animals to be slaughtered to avoid unnecessary suffering, but the Intervention Board that administers the scheme has only managed to slaughter 35,300 animals since the scheme started on 23 March. Its officials admitted this week that 98,531 animals were added to the waiting list on the same day that only 3,304 were slaughtered.

I am amazed that the RSPCA isn't concerned at so much suffering being permitted as a result of civil-service incompetence. The least that I expect from the RSPCA is a high-profile campaign highlighting the problem.

If Robert Burns had been alive today he would certainly have written some scathing verses about the RSPCA turning a blind eye to the plight of all those newborn lambs, piglets and calves that are needlessly suffering an agonising, slow death.

Thursday, 5 April

Some farmers are worried that after banking their compensation cheques the banks will not allow them the same overdraft facilities when they come to restock.

That's the difference between banker and terrorists. At least you can negotiate with a terrorist.

Friday, 6 April

The SNFU announces that it is putting up £150,000 to help finance a meat-promotion campaign through the recently formed Quality Meat Scotland.

It's only right that farmers should be the first to put their hands in their pockets in the fight to win back all those lost markets.

Saturday, 7 April

Two cases confirmed in Jedburgh. It's frightening the way the disease can suddenly flare up in new areas.

Sunday, 8 April

An outbreak confirmed in Wigtownshire. The unfortunate farmers involved have several farms, so this case is potentially devastating for the area with its concentration of large dairy herds.

Nick Brown makes a plea for farmers not to appeal against the slaughter of their stock as it holds up the process of containing the disease.

Monday, 9 April

The Intervention Board finally gets the welfare cull underway. It's ridiculous it has taken them so long.

Tuesday, 10 April

Farmers themselves are spreading the disease by human contact and illegally moving livestock. The police are currently investigating 350 suspected cases of illegal movements. The Scottish National Farmers' Union appeals for continued vigilance and cites the following examples of situations where the disease was inadvertently passed on:

A worker on a farm with no signs of disease went home and fed his sheep. The first farm turned out to be incubating the disease so both farms went down with it.

Two farms are in the same ownership and some distance apart. One has a Longtown connection. Workers and machinery move between the two so both farms get the disease.

A farmer does relief milking on premises with no sign of disease, but foot-and-mouth was incubating so his farm went down.

A feed wagon was used to feed silage on two farms. The first farm later showed signs of the disease, then the second.

Wednesday, 11 April

Experts are suggesting that the disease may have peaked.

Thursday, 12 April

A Scottish farmer has slaughtered his entire flock of field ewes and about a quarter of his beef cows under the welfare scheme. Apparently he claimed that the restriction on the movement of livestock had prevented him from selling stock and that would lead to poorer grazing during the summer to the detriment of the animals' welfare.

Many reckon he could have bought extra feed or fertilisers and are speculating that he has blatantly milked the system to get the attractive compensation for low-value stock. I am disgusted that a farmer is capable of needlessly slaughtering newly born, healthy lambs and calves. I also wonder how such unjustified destruction is possible.

Animals should only be slaughtered under the scheme to prevent unnecessary suffering and not to line pockets at the taxpayer's expense.

Saturday, 14 April

Vaccination is back on the agenda for Cumbria and Devon.

Jim Walker, the SNFU president, warns that if vaccination does get the go ahead in England then we will have to reinstate the border between England and Scotland as that will be the only way that we can preserve our foot-and-mouth-free status for exports.

The SNP are bound to be cock-a-hoop with that idea and offering our Jim free party membership! I must check that my passport is still valid.

Sunday, 15 April

Brian Pack – the chief executive of the mighty Aberdeen and Northern Marts group, Europe's biggest farmer co-operative – warns that the only hope for the lamb market this season is regional status for export for provisionally free areas under foot-and-mouth controls.

According to him, 'The number of livestock being slaughtered is horrendous, but when measured against the number of animals we exported it's irrelevant.' He added: 'Our problems are major, there must be 30 per cent or so of lambs not required and there's no way on earth we need those in the UK.'

That sounds like a powerful argument against vaccination.

Monday, 16 April

Although there will probably be a massive glut of lambs in the autumn, those with early spring lambs to sell could be in for a bonanza. Such lambs tend to be produced in the hardest-hit areas like Devon and Dumfries and Galloway, and a fair proportion have already been culled.

Wednesday, 18 April

A Dumfriesshire farm near Auldgirth had the disease confirmed after cattle recently turned out to grass developed the symptoms. The farm had been thoroughly inspected by vets a month ago and tests on the sheep had proved negative after the farm was identified as having had an extremely dangerous contact with a livestock dealer who had infected animals.

That's an extremely worrying development and raises doubts over the test. It could be a coincidence and the sheep may have been disease-free and subsequently became infected. Then again, if the testing process is unreliable the implications for other areas where tests have been negative are horrendous.

I am also mindful of recent comments by a local vet who reckons the disease is now endemic in 80 per cent of the hill flocks in Dumfries and Galloway.

We will know for definite within the next fortnight as hill lambings get underway and cattle are turned out to grass, as that's when the disease will flare up if it has been harboured unnoticed in adult sheep.

Thursday, 19 April

Fifty Devonshire vets say that slaughtering cattle is madness. I reckon that keeping their views to themselves till now is also madness.

Farming leaders reject Tony Blair's notion for vaccination. It's probably now too late in the day anyway.

Friday, 20 April

Leading government-scientific-adviser, professor David King, announces that he is confident that the foot-and-mouth epidemic is under control. I seem to remember Nick Brown making a similar statement in the Commons over a month ago.

Sunday, 22 April

There is growing concern at health risks from funeral pyres when government research reveals that there have been high levels of pollution from cancer-causing dioxins. Scientists reckon that the pyres are now emitting more dioxin than the whole of British industry.

Ian Anderson, director of operations for the Scottish Executive, said it was, 'Not safe to draw conclusions from what is happening in England. We use different materials like coal and woodchips instead of tyres and the other materials being used down south.' Some reckon that napalm would make a better job of incinerating the carcasses.

A slaughterman in Cumbria is suspected of contracting foot-and-mouth.

Monday, 23 April

Two more people suspected of having contracted the disease. I suppose it's only a matter of time before some obscure rent-a-quote professor starts another food scare by predicting an epidemic of human foot-and-mouth.

Tuesday, 24 April

Swallows return to my farm buildings.

There are real doubts about the validity of foot-and-mouth testing procedures. The blood test is for antibodies to the virus. However, antibodies will not appear in the blood for at least four days after infection. Antibody levels peak after about one week from infection, remain high for the second week and then fall off. In mild cases the number of antibodies reduces and a small number of infected animals produce none.

There is also a tissue test for skin covering a blister on the animal. It is difficult to take and if any disinfectant (either from the vet or the animal) comes into contact, it kills the virus. There is only a window of about a week in the infection cycle for this test to work.

Wednesday, 25 April

Fears are growing that foot-and-mouth may have got into the wild-deer population. Although ministry laboratory tests have been unable to detect it so far, there are worries that it could establish a permanent infective reservoir. According to Dr John Fletcher, past president of the Veterinary Deer Society, 'There have been marked clinical signs in roe deer in Cumbria, Dumfriesshire and Devon, but they have all come back negative.'

Thursday, 26 April

Phoenix the calf is spared when the government announces a relaxation of the slaughter policy. I suspect the plight of a new-born calf was too emotive a subject in the run up to an election.

It seems only yesterday that the situation was so dire we had to consider vaccination. I am uneasy about policy being changed on the hoof and not at all happy about greater dependence on unreliable testing.

Friday, 27 April

A survey by *Farmers Weekly* reveals that half of the farmers affected by the disease plan to quit farming, or scale down their business after the crisis.

I know how they feel!

As it turned out I was lucky enough not to get the disease on my farm, although I suffered serious financial losses as a consequence of the epidemic. The last case in Scotland was at the end of May 2001 and the last case in the United Kingdom was at the end of September. During the epidemic 4.2 million animals were slaughtered on 1,051 farms. In addition, there was a welfare cull of just under a million sheep, 228,000 pigs, 139,500 cattle and 1,365 goats giving a total killed of 5.57 million. The total cost to the taxpayer was more than £8 billion and many rural businesses have never fully recovered.

I hope I never see the like again!